T0324469

LEARNING EXPERIENCES TO PROMOTE MATHEMATICS LEARNING

Yearbook 2014
Association of Mathematics Educators

LEARNING EXPERIENCES TO PROMOTE MATHEMATICS LEARNING

Yearbook 2014
Association of Mathematics Educators

editors

Pee Choon Toh
Tin Lam Toh
Berinderjeet Kaur

Nanyang Technological University, Singapore

 World Scientific

 AME
ASSOCIATION
OF MATHEMATICS
EDUCATORS

Published by

World Scientific Publishing Co. Pte. Ltd.

5 Toh Tuck Link, Singapore 596224

USA office: 27 Warren Street, Suite 401-402, Hackensack, NJ 07601

UK office: 57 Shelton Street, Covent Garden, London WC2H 9HE

British Library Cataloguing-in-Publication Data
A catalogue record for this book is available from the British Library.

LEARNING EXPERIENCES TO PROMOTE MATHEMATICS LEARNING
AME Yearbook 2014

ISBN 978-981-4612-90-6

Printed in Singapore

Contents

Chapter 1

It Matters How Students Learn Mathematics

Berinderjeet KAUR

This introductory chapter details the emphasis of the 2012 school mathematics curriculum in Singapore and also provides an overview of the chapters in the book. The chapters centre around three main areas: fundamentals for active and motivated learning, learning experiences for developing mathematical processes, and the use of ICT tools for learning. It ends with some concluding thoughts that readers may want to be cognizant of while reading the book and also using it for reference and further work.

1 Introduction

This yearbook of the Association of Mathematics Educators (AME) in Singapore focuses on Learning Experiences to Promote Mathematics Learning. Like all of the past yearbooks, Mathematical Problem Solving (Kaur, Yeap, & Kapur, 2009), Mathematical Applications and Modelling (Kaur & Dindyal, 2010), Assessment in the Mathematics Classroom (Kaur & Wong, 2011), Reasoning, Communication and Connections in Mathematics (Kaur & Toh, 2012) and Nurturing Reflective Leaners in Mathematics (Kaur, 2013), the theme of this book is also shaped by the foci of the school mathematics curriculum developed by the Ministry of Education (MOE) and the needs of mathematics teachers in Singapore schools.

The 2012 syllabus for mathematics in Singapore schools (MOE, 2012a; 2012b) places heightened emphasis on the role of learning experiences for mathematics learning. It states that:

Learning mathematics is more than just learning concepts and skills. Equally important are the cognitive and metacognitive process skills. These processes are learned through carefully constructed learning experiences. For example, to encourage students to be inquisitive, the learning experiences must include opportunities where students discover mathematical results on their own. To support the development of collaborative and communication skills, students must be given opportunities to work together on a problem and present their ideas using appropriate mathematical language and methods. To develop habits of self-directed learning, students must be given opportunities to set learning goals and work towards them purposefully. A classroom rich with these opportunities, will provide the platform for students to develop 21st century competencies (MOE, 2012a, p. 20; 2012b, p. 20).

2 Learning Experiences in the Mathematics Syllabuses

In the 2012 mathematics syllabuses for schools in Singapore "learning experiences are stated in the mathematics syllabuses to influence the ways teachers teach and students learn so that the curriculum objectives can be achieved" (MOE, 2012a, p. 20; 2012b, p. 20). Figure 1 shows an excerpt from the primary school syllabus and Figure 2 shows an excerpt from the secondary school syllabus. From both figures, it is apparent that statements expressed in the form "students should have the opportunities to ..." focus teachers on the student-centric aspect of learning mathematics. The statements describe actions that would allow students to i) engage in co-creation of knowledge ii) make sense of the knowledge they acquired and iii) work collaboratively and communicate their reasoning using mathematical vocabulary.

The syllabuses, an intended curriculum, serve as a guide for teachers to undertake school-based curriculum innovations and customisations in line with the broad aims of mathematics education in Singapore. The aims are to acquire and apply mathematical concepts and skills; develop cognitive and metacognitive skills through a mathematical approach to problem solving; and develop positive attitudes towards mathematics

(MOE, 2012a, p. 7; 2012b, p. 7).

Primary Two	
Sub-Strand: Whole Numbers	
Content	Learning Experience
1. Numbers up to 1000	Students should have opportunities to:

1.1 counting in tens/hundreds
1.2 number notation, representations and place values (hundreds, tens, ones)
1.3 reading and writing numbers in numerals and in words
1.4 comparing and ordering numbers
1.5 patterns in number sequences
1.6 odd and even numbers
	a) give examples of numbers in everyday situations, and talk about how and why the numbers are used.
	b) work in groups using concrete objects/the base-ten set/play money to
	  - count in tens/hundreds to establish 10 tens make 1 hundred and 10 hundreds make 1 thousand.
	  - represent and compare numbers.
	c) make sense of the size of 100 and use it to estimate the number of objects in the size of hundreds.
	d) use the base-ten set/play money to represent a number that is 1, 10 or 100 more than/less than a 3-digit number.
	e) use place-value cards to illustrate and explain place values, e.g. the digit 3 stands for 300, 30 or 3 depending on where it appears in a number.
	f) use place-value cards to compare numbers digit by digit from left to right, and use language such as 'greater than', 'greatest', 'smaller than', 'smaller than', 'smallest' and 'the same as' to describe the comparison.
	g) describe a given number pattern before continuing the pattern or finding the missing number(s).

2. Addition and Subtraction	Students should have opportunities to:

2.1 addition and subtraction algorithms (up to 3 digits)
2.2 solving up to 2-step word problems involving addition and subtraction
2.3 mental calculation involving addition and subtraction of a 3-digit number and ones, tens, hundreds
	a) write addition and subtraction equations for number stories and explain the meaning of the equal sign.
	b) achieve mastery of basic addition and subtraction facts within 20 by
	  - writing a family of 4 basic addition and subtraction within given any one of the basic facts (e.g. 9+7=16, 7+9=16, 16-9=7, 16-7=9 are a family of addition and subtraction facts).
	  - playing games, including applets and digital games.
	c) work in groups using the base-ten set/play money to illustrate the standard algorithms for addition and subtraction up to 3 digits.
	d) use the part-whole and comparison models to illustrate the concepts of addition and subtraction and use the models to determine which operation (addition or subtraction) to use when solving 1-step word problems.
	e) use the comparison model to reinforce the language of comparison such as "Ali has 30 more stickers than Siti."

Figure 1: An excerpt from the primary mathematics syllabus (MOE, 2012a, p. 37)

Secondary Three/Four (O-Level Mathematics)	
Strand: Geometry and Measurement	
Content	Learning Experience
G2 Congruency and similarity	Students should have opportunities to:

2.7 determining whether two triangles are
  - congruent
  - similar
2.8 ratio of areas of similar plane figures
2.9 ratio of volumes of similar solids
	a) construct triangles with given conditions, e.g. "3 sides", "3 angles", "2 sides, 1 angle", and '2 angles, 1 side", and examine what conditions are necessary for congruency/similarity.

G3 Properties of Circles	Students should have opportunities to:

3.1 symmetry properties of circles
  - equal chords are equidistant from the centre
  - the perpendicular bisector of a chord passes through the centre
  - tangents from an external point are equal in length
  - the line joining an external point to the centre of the circle bisects the angle between the tangents
3.2 angle properties of circles
  - angle in a semicircle is a right angle
  - angle between tangent and radius of a circle is a right angle
  - angle at the centre is twice the angle at the circumference
  - angles in the same segment are equal
  - angles in opposite segments are supplementary
	a) use paper folding to visualise symmetric properties of circles, e.g. "the perpendicular bisector of a chord passes through the centre".
	b) Use GSP or other dynamic geometry software to explore the properties of circles, and use geometrical terms correctly for effective communication.

G4 Pythagoras' theorem and trigonometry	
4.4 extending sine and cosine to obtuse angles 4.5 use of the formula $\frac{1}{2} ab \sin C$ for the area of a triangle 4.6 use of sine rule and cosine rule for any triangle 4.7 problems in two or three dimensions including those involving angles of elevation and depression and bearings	a) visualise the height, north direction, right-angled triangle, etc. from 2D drawings of 3D situations. b) use the sine and cosine rules to articulate the relationships between the sides and angles of a triangle, e.g. the lengths of the sides are proportional to sine of the corresponding angles, Pythagoras theorem is a special case of the cosine rule, etc.
G5 Mensuration	
5.7 arc length, sector area and area of a segment of a circle 5.8 use of radian measure of angle (including conversion between radians and degrees)	a) find the arc length and sector area by considering them as fractions of the circumference and area of circle respectively. b) visualise the size of an angle of 1 radian, and estimate the size of angles in radians.

Figure 2: An excerpt from the secondary mathematics syllabus (MOE, 2012b, pp. 57-58)

In addition the syllabus documents outline three principles of mathematics teaching and three phases of mathematics learning in the classrooms. The three principles of teaching are as follows:

Principle 1 – Teaching is for learning; *learning is for understanding*; understanding is for reasoning and applying and, ultimately problem solving.

Principle 2 – Teaching should build on students' knowledge; take cognizance of students' interests and experiences; and engage them in *active* and *reflective learning*.

Principle 3 – Teaching should connect learning to the real world, harness ICT tools and emphasise 21st century competencies (MOE, 2012a, p. 23; 2012b, p. 21).

The three phases of mathematics learning in the classrooms are as follows:

Phase I – Readiness
Student readiness to learn is vital to learning success. In the readiness phase of learning, teachers prepare students so that they are ready to learn. This requires considerations of prior knowledge, *motivating contexts*, and learning environment.

Phase II – Engagement
This is the main phase of learning where teachers use a repertoire of pedagogies to engage students in learning new concepts and skills. Three pedagogical approaches, *activity-based learning*, *teacher-directed inquiry*, and direct instruction, form the spine that supports most of the mathematics instruction in the classroom. They are not mutually exclusive and could be used in different parts of a lesson or unit. For example, the lesson or unit could start with an activity, followed by teacher-led inquiry and end with direct instruction.

Phase III – Mastery
This is the final phase of learning where teachers help students consolidate and extend their learning. The mastery approaches include: ***motivated practice***, reflective review and extended learning. (MOE, 2012a, p. 24-27; 2012b, pp. 22-25).

The learning principles and phases of learning in the Singapore school mathematics syllabus documents signal very clearly the need for teachers to engage their students in active learning and learning for understanding in contexts that are motivating and help them develop 21st century competencies. To address the challenge faced by teachers, in Singapore schools, in harnessing appropriate learning experiences to engage students in their learning of mathematics the theme of the 2013 conference for teachers, jointly organised by the Association of Mathematics Educators (AME) and the Singapore Mathematical Society (SMS), was appropriately *Learning Experiences to Promote Mathematics Learning*.

The following 13 peer-reviewed chapters resulted from the keynote lectures delivered and invited workshops conducted during the conference. The authors of the chapters were asked to focus on evidence-based practices that school teachers can experiment in their lessons to bring about meaningful learning outcomes. The chapters centre around three main areas, namely fundamentals for active and motivated learning, learning experiences for developing mathematical processes and use of ICT tools for learning through visualisations, simulations and representations. It must be noted that the 13 chapters do give a reader some ideas about meaningful learning experiences that promote mathematics learning. However, in no way are the 13 chapters a collection of all the know-how of the subject.

3 Fundamentals for Active and Motivated Learning

Learning results from students' active participation in activities that provide learning experiences. In the classroom, teachers guide students in their learning through instructional activities. Often teacher's knowledge and beliefs guide them in shaping the activities. So, for designing learning experiences that are relevant and engaging, some necessary fundamentals are explored in this book. Wong in chapter 2, notes that motivating students to learn mathematics with active engagement is a tough challenge for some teachers, especially so when the topics do not have obvious daily applications due to their abstract nature. He draws on theories and research in the literature related to motivation and derives a framework of motivation to learn called *M_Crest*. The framework covers six motivators: *M*eaningfulness, *C*onfidence, *R*elevance, *E*njoyment, *S*ocial relationships, and *T*argets. In the chapter mathematical examples are provided to illustrate the implementation of these motivators. Suggestions for teacher research to investigate the effects of the motivators are also discussed.

In chapter 3, Toh draws on the works of Taba (1962) and Tyler (1946) in the fields of curriculum design and development. He illustrates how each of Tyler's five principles for selecting appropriate learning experiences may be drawn upon when designing learning experiences for effective instruction. He uses examples from secondary mathematics. Kwon, Park and Park, in chapter 4, note that learning by doing is an effective method of learning and authentic learning is a proponent of such a method. Drawing on Herrington and Herrington's (2006) principles for designing authentic learning environments, they show how teachers may provide students with authentic learning experiences through 3 D printing technology.

Although research has shown that Pedagogical Content Knowledge (PCK) of teachers did predict student achievement, it was also found that teacher's PCK is dependent on their Mathematics Content Knowledge (MCK) (Baumert et al., 2010). Beswick in chapter 5 uses items, measuring MCK and PCK, from an on-line survey to illustrate how these items may be used for professional learning and the kinds of knowledge teachers might need in order to use these sorts of items to create learning experiences that promote meaningful mathematics learning.

4 Learning Experiences for Developing Mathematical Processes

The seven chapters in this area provide teachers with ample examples of learning experiences that lead to the development of mathematical processes: thinking and reasoning skills. According to Fawcett (1938/1995) definitions and propositions socially constructed by students and their teacher facilitate "critical thinking". In chapter 6, Shimizu illustrates how defining, extending and creating mathematical relationships offer students rich and meaningful learning experiences. He draws on a teaching experiment in Japan with grade 10 students on the topic of kite and "boomerang" and demonstrates how teachers may engage students in examining a definition, extending it and creating new ones. Cheng, in chapter 7, draws on the four teaching phases, developed by White and Mitchelmore (2010), and shows how mathematical abstraction may be facilitated in primary 5 classrooms. She also affirms that student-centred activities offer students with opportunities to be pro-active in their learning. They also enjoy working in groups, and communicating their ideas. The sequence of three activities, in the appendix to the chapter, provides teachers with an example of a learning experience for abstraction developing the concept percentage.

In the primary levels, learning what numbers mean, how they may be represented, relationships amongst them and computing with them are keys to developing number sense. McIntosh, Reys and Reys (1992) developed a framework comprising three broad categories and identified six related strands that provide a structure for designing learning experiences that facilitate the development of number sense. In chapter 8, Yeo illustrates how each of the strands may be developed in primary mathematics lessons through learning experiences which are outlined in the chapter. Hodgen, Küchemann and Brown, in chapter 9, note that when the teaching of algebra emphasises procedural manipulation of symbols over a more conceptual understanding learners see it as a system of arbitrary rules. Drawing on their experiences from the Increasing Competence and Confidence in Algebra and Multiplicative Structures (ICCAMS) study and they show how learning experiences may be planned to develop algebraic thinking amongst learners.

In chapter 10, Brown, Hodgen and Küchemann note that successful progress in learning mathematics depends on a sound foundation of the understanding of multiplicative structures and reasoning. This includes not only the properties and meanings of multiplication and division, but also their many links with ratio and percentage and with rational numbers - both fractions and decimals. As primary teaching can sometimes emphasise facility in calculation rather than the building of conceptual connections, these connections which take time to establish are neglected. This chapter draws on their experiences from the (ICCAMS) study and discusses how learning experiences can be planned to develop multiplicative reasoning using models to foster understanding. Mathematical induction (MI) is a technique that high school students may need when constructing proofs for mathematical assertions. In chapter 11, Tay and Toh, illustrate how the pedagogy of the technique of MI may be set within the natural environment of problem solving. They show how teachers may teach MI while engaging students in authentic problem solving. Zhao, in chapter 12, laments that the teaching of mathematical proofs lacks adequate emphasis in the A-Level curriculum for Singapore schools. He shows how appropriate scaffolding can be crafted to guide students in learning mathematical proofs via mathematical problem solving; and how proof problems can be constructed for use in teaching A-Level Mathematics. He also appends a list of proof questions, to the chapter, which can be used by teachers for instructional needs.

5 Use of ICT Tools for Learning

Information and communications technology (ICT), increasingly available at both home and school, offers new kinds of learning experiences for students. Kissane, in chapter 13, explores the distinctive contribution of ICT to learning experiences of students, focusing on those that are not readily available through traditional media of pencil and paper and organised practice of skills. The chapter highlights examples that use virtual manipulatives on the Internet, other web-based experiences, and hand-held devices such as Apple's iPad® and iPod Touch® and modern scientific calculators. The focus of the examples is on the development and understanding of important ideas in the number curriculum, such as those related to number patterns, place value, decimals, fractions, factors and ratios, rather than to the refinement of arithmetical skills.

In the last chapter of the book, Kemp notes that ICT offers different kinds of experiences for learning mathematics from those found in textbooks and traditional teaching in the classroom. In the chapter Kemp examines how features of ICT can be used to help create mathematical meaning for secondary students, inside and outside the classroom in the areas of algebra and geometry. In both the chapters, Kissane and Kemp, emphasise that learning experiences using ICT demand a new kind of attention to the role of the teacher, in a variety of contexts, including individual ICT use, small-group use and whole-class use.

6 Some Concluding Thoughts

Teacher exposition and excessive practice of algorithms in mathematics classrooms have resulted in many students leaving school with a collection of well-practiced procedures and formulae but with only a vague grasp of their meaning or of when they might be used. In my work with parents when attempting to distinguish conceptual and procedural understanding I often ask the question: What is area? The answer I almost always get from them is "length times breadth", and when I hesitate to say if it is correct or wrong they go on to say it must be "half base times height !".

The need to learn by doing and reflecting on the activity is what John Dewey advocated more than a century ago (Kilpatrick and Silver (2000, p. 226). Therefore the role of learning experiences in learning mathematics is not new. It matters how students learn mathematics as the experience shapes their perception and use of the knowledge. But the challenge for teachers in Singapore schools is striking a balance between engaging their students in meaningful learning and also preparing them for examinations. For this to occur, teachers need to move away from their main preoccupation with drill and practice as observed by Wong and Lee (2010):

Drilling students to solve standard problems is a major part of mathematics instruction in Singapore and many other countries.

Singapore teachers often construct exercises that parallel past examination questions, which have been compiled and sold to the public as Ten-Year Examination Series. Even problems with initially unfamiliar contexts or "challenge problems" gradually become routine to students with frequent exposure (p. 102).

The chapters in the yearbook provide readers with some ideas on the why, what and how of learning experiences for the learning of mathematics. Readers are urged to read the chapters carefully and try some of the ideas in their classrooms and convince themselves that these ideas offer a means of infusing learning experiences in their lessons and engage students in meaningful mathematical practices.

References

Baumert, J., Kunter, M., Blum, W., Brunner, M., Voss, T., Jordan, A., & Tsai, Y.-M. (2010). Teachers' mathematical knowledge, cognitive activation in the classroom, and student progress. *American Educational Research Journal, 47*(1), 133-180.

Fawcett, H.P. (1938/1995). *The nature of proof: Description and evaluation of certain procedures used in a senior high school to develop an understanding of the nature of proof.* New York: Columbia University Teachers College Bureau of Publications. (Reprinted by National Council of Teachers of Mathematics, Reston, VA: The Council)

Herrington, A., & Herrington, J. (2006). What is an authentic learning environment? In A. Herrington, & J. Herrington (Eds.), *Authentic learning environments* (pp. 1-13). Hershey, PA: ISP.

Kaur, B. (2013). *Nurturing reflective learners in mathematics.* Singapore: World Scientific.

Kaur, B., & Dindyal, J. (2010). *Mathematical applications and modelling.* Singapore: World Scientific.

Kaur, B., & Wong, K.Y. (2011). *Assessment in the mathematics classroom.* Singapore: World Scientific.

Kaur, B., & Toh, T.L. (2012). *Reasoning, communication and connections in mathematics.* Singapore: World Scientific.

Kaur, B., Yeap, B.H., & Kapur, M. (2009). *Mathematical problem solving.* Singapore: World Scientific.

Kilpatrick, J., & Silver, E.A. (2000). Unfinished business: Challenges for mathematics educators in the next decades. In, M.J. Burke & F.R. Curio (Eds.), *Learning mathematics for a new century* (pp. 223-235). Reston, VA: National Council of Teachers of Mathematics.

McIntosh, A., Reys, B.J., & Reys, R.E. (1992). A proposed framework for examining basic number sense. *For the Learning of Mathematics*, 12(3), 2-8. 329.

Ministry of Education, Singapore (2012a). *Primary mathematics teaching and learning syllabus.* Singapore: Author.

Ministry of Education, Singapore (2012b). *O-Level, N(A) Level, N(T) level mathematics teaching and learning syllabuses.* Singapore: Author.

Taba, H. (1962). *Curriculum development: Theory and practice.* New York: Harcourt Brace and World.

Tyler, R.W. (1946). *Basic principles of curriculum and instruction.* Chicago: The University of Chicago Press.

White, P., & Mitchelmore, M.C. (2010). Teaching for abstraction: A model. *Mathematical Thinking and Learning*, *12*, 205-226.

Wong, K.Y., & Lee, N.H. (2010). Issues of Singapore mathematics education. In, K.S. F. Leung & Y. Li (Eds.), *Reforms and issues in school mathematics in East Asia* (pp. 91-108). Rotterdam, The Netherlands: Sense.

Chapter 2

M_Crest: A Framework of Motivation to Learn Mathematics

WONG Khoon Yoong

Motivating students to learn mathematics with active engagement is a tough challenge for some teachers. This is especially so for teaching mathematics in secondary schools, because many topics at this level do not have obvious daily applications due to their abstract nature. This paper explains how theories and research in the motivation literature can be used to derive a new framework of motivation to learn mathematics called *M_Crest*. This framework covers six motivators: *Meaningfulness, Confidence, Relevance, Enjoyment, Social relationships,* and *Targets*. Mathematical examples are given to illustrate the implementation of these motivators. Suggestions for teacher research to investigate the effects of these motivators are also discussed.

1 Introduction

Students come to lessons with diverse life experiences, cognitive learning tools, emotional propensities, and social skills. Their motivation to learn any subject can be conceptualised as the extent to which they are willing to invest these cognitive, affective, and social resources to complete the given learning tasks. Almost every teacher has encountered students who are not even mildly motivated to study mathematics. There are many reasons why secondary students in particular are not motivated or do not exert adequate effort into learning mathematics. These include boring lessons, tasks that are too challenging or too easy, punishments for poor quality work, poor rapport with the teacher and classmates, lack of relevance of many abstract mathematics topics to their daily life, and so forth. Searching for ways to motivate these secondary school students to learn mathematics is an ongoing challenge for many teachers.

In this chapter, a new framework called *M_Crest* is proposed to help teachers in their continual search for viable motivation strategies, although there are no silver bullets that can solve all the motivation woes in the classrooms. This framework covers these six influential motivators: *Meaningfulness, Confidence, Relevance, Enjoyment, Social relationships,* and *Targets*. These motivators are derived from the vast and complex field of generic motivation theories (see below), practical applications of motivation theories to mathematics education (e.g., Brahier & Speer, 2011; Middleton & Jansen, 2011; Posamentier & Krulik, 2012), insights from international comparative studies, and recommendations given by the Singapore Ministry of Education.

2 Motivation Theories

According to the online Oxford Dictionaries[1], the word *motive* originated from the Latin word *movere,* meaning "to move", so that to *motivate* someone is to provide reasons to move the person to do something. On the basis of these two ideas of *reason* and *move,* a vast and complex field about motivation has emerged (e.g., Pink, 2009; Sansone & Harackiewicz, 2000; Schunk, Pintrich, & Meece, 2008; Stipek, 2002; Woolfolk, Hughes, & Walkup, 2013), resulting in many theories about motivation in education. These theories can be placed into the six categories shown in Table 1. Affective theories focus on internal emotions as the main source of motivation, behavioural theories on extrinsic rewards and punishment, cognitive theories on meaningfulness and achievement, socio-cultural theories on needs of belonging to social and cultural groups, humanistic theories on self-actualisation and values, and neuro-scientific theories on the physiological basis of emotions and

---

[1] http://oxforddictionaries.com/

social cognition. Thus, motivation is a multi-dimensional construct that overlaps with constructs such as drive, self-regulation, arousal, needs, autonomy, goal orientations, attribution, and self-efficacy. This complexity provides rich interplay of constructs but it also leads to a lack of clarity to guide practices for practitioners and to summarise research findings based on different meanings attributed to motivation. While these problems cannot be readily resolved, what is offered here is a personal synthesis in the form of the *M_Crest* framework.

Table 1
Motivation theories and words offered by teachers

Theories	Words Offered by Teachers
Affective	interested, passion, happy, inspired, confident, love maths; *troubled, bored*
Behavioural	persistent, self-driven, independent, engaged, discipline, proactive; *day dreaming, disruptive, distracted, asleep, slouching, talkative, lazy, passive*
Cognitive	seek to clarify, inquisitive, ask thought provoking questions, ask for challenging questions, seek alternative solutions, reflective
Socio-cultural	respect teacher
Humanistic	(none)
Neuro-scientific	(none)

At a recent workshop, secondary school teachers were asked to think of words that they associated with students who were motivated or demotivated to learn mathematics. Some of these words are shown in the last column of Table 1; the demotivated words are in italics. Note that the words grouped under *behavioural* are *not* about rewards and punishments as intended by behaviourist theories of motivation; instead, they describe potential behaviours of motivated or demotivated students. Although the grouping of these words is not precise, it is apparent that to these teachers, motivation is related to the affective, behavioural, and cognitive aspects, in almost equal emphasis. On the other hand, demotivation is predominantly about behaviours, somewhat related to affect, and none about cognition and the other three categories. It seems that these teachers were looking for demotivation in terms of classroom behaviours, which are observable, rather than underlying attributes, which are more difficult to identify. Furthermore, these teachers were not aware of motivation in terms of self-experiences and values (humanistic) or physiological-neurological factors. Of course, this brief activity does not constitute serious research about what Singapore secondary school teachers know about motivation, yet the findings do point to areas for further consideration. For instance, many books and articles have been published in recent years about brain-based learning (e.g., Caine et al., 2009; de Jong et al., 2009; Sousa, 2008), providing evidence that there are measurable changes in the brain as a result of engaging in different types of learning. Some findings from neuroscience will be noted in later sections.

3 The *M_Crest* Framework

Numerous motivators can be culled from the theories mentioned above, and, for practical purposes, it is necessary to choose only a few impactful motivators so that the resulting scheme is manageable. Six items are easy to remember, and placing them in a memorable acronym helps recall of the intended motivators. In this case, the word *crest* with its meaning of being at the *top* or *summit* emphasises that teachers should help *all* their students to be maximally motivated to do the best in order to achieve *peak* performance in learning. Which six motivators to be selected certainly depends on the person making the choice, and what have been selected for the *M_Crest* framework are based on the author's personal experiences as a mathematics educator for the past four decades working in four different countries. Four out of the six motivators are mentioned in the Singapore mathematics curriculum (Ministry of Education, 2012a):

> Making the learning of mathematics *fun*, *meaningful* and *relevant* goes a long way to inculcating positive attitudes towards the subject. Care and attention should be given to the design of learning activities,

to build *confidence* in and develop appreciation for the subject. (p. 9; italics added to the four relevant words)

3.1 *M = Meaningfulness*

The *M_Crest* framework stresses that the most crucial motivator is that students should be able to make sense of the mathematics they are studying. When students understand conceptually, procedurally, and relationally what they are supposed to learn, they are more likely to be motivated enough to put in the effort to complete the learning activities. Unfortunately, it is well-documented that many students do not find abstract mathematics meaningful, and they have to resort to memorising many disjointed rules, which are easily forgotten and prone to errors. Thus, they become demotivated in their learning.

A major contribution to this sad affair is inappropriate teaching, in particular, when teachers use only words and symbols to explain mathematical concepts, rules, and proofs. Even well-known people have suffered under such poor teaching. For example, the eminent British philosopher-cum-mathematician Bertrand Russell (1872 – 1970) recounted the following incidence when he was learning algebraic expansion as a boy:

> I was made to learn by heart: "The square of the sum of two numbers is equal to the sum of their squares increased by twice their product". I had not the vaguest idea of what this meant, and when I could not remember the words, my tutor threw the book at my head, which did not stimulate my intellect in any way. (Russell, 1975, p. 31)

Obviously Russell recovered from this bad teaching and he even "enjoyed impressing a new tutor with my knowledge" (ibid). The psychiatrist Carl Jung (1875 – 1961), a contemporary of Russell, faced a similar problem:

Mathematics classes became sheer terror and torture to me. I was so intimidated by my incomprehension [of algebra] that I did not dare to ask any questions[2].

Such incomprehension that arises from poor explanation and fear to ask questions can be avoided with lessons that emphasise meanings and understanding. Two techniques are offered below: the use of patterns and variations in different modes of representation of mathematical objects.

Patterns

Mathematics is a study of patterns, and the search for surprising patterns can be very motivating. Many mathematics patterns are extensions from one domain to another in order to preserve certain properties. For example, mathematicians give meaning to the expression a^0 ($a \neq 0$) by examining what happens if one keeps dividing a^n by a in the case of natural numbers (n) so that the laws of indices remain applicable when n is zero or negative. Using patterns to develop concepts and skills is powerful because they are easier to understand compared to mere words and symbols and they build on prior knowledge, which is a very powerful learning principle supported by theory and research (Ausubel, 1968; Hattie, 2009).

Another example of the use of patterns in teaching concerns the addition and subtraction of integers, which is quite problematic for many lower secondary students. Addition of a positive integer is often introduced as moving to the right side of a number line and of a negative integer as moving to the left. The movements are changed for subtraction. However, it is not often pointed out to the students how the two types of movements are inter-related. Figure 1 is an attempt to show this link. Subtracting a positive integer involves the same movement as adding a negative integer; this reinforces the ideas of subtraction as the opposite (reverse) operation to addition and negative as "opposite" to positive. The same pattern applies to subtracting a negative integer and adding the positive one. Making these patterns explicit helps students to make better sense of why the movements are chosen in that particular

---

[2] http://www.famousquotesabout.com/quote/The-teacher-pretended-that/21652

way, so that they come to appreciate that mathematical rules are not mindless and arbitrary "tricks".

Operation	Second number	
	Positive	Negative
Addition	Move to right →	*Move to left ←*
Subtraction	*Move to left ←*	Move to right →

Figure 1. Links between addition and subtraction on a number line

Variations in different modes of representation

Verbal explanation is only one mode of teaching, and the experiences of Russell and Jung mentioned above show that relying solely on words to explain abstract mathematics is neither effective nor motivating. Words build on more words and this approach is quite removed from concrete experiences that are essential for concept development among students who are still in the concrete operational stage. The perceptual variability principles and the mathematical variability principle expounded by Dienes (1964) can be used to design diverse learning experiences to make mathematics meaningful to the students.

Consider the justification of the formula for the area of a parallelogram as base × height. A typical activity is to get students to draw the parallelogram in Figure 2(a), cut off the triangle and move it to the right in order to make a rectangle. This hands-on activity is certainly better than verbal explanations to help students "see" why the formula works. The visual element is also helpful.

(a)   (b)

Figure 2. Different forms of a parallelogram

Dienes' principles suggest that this approach can be further extended by changing some elements of the parallelogram, for example, what happens if the perpendicular from the top left vertex "falls" outside the base, as shown in Figure 2(b)? How can the parallelogram be cut to show that the formula still works? In how many different ways can this be done? Opening up a standard formula in this way to different variations will motivate students to explore the conditions under which the formula may be applicable, thereby enhancing the meaningfulness of the rule.

Finally, the activities used to explore this formula can be put together on a multi-modal thinkboard, which is partially completed, as shown in Figure 3. The thinkboard can promote meaningfulness of mathematics by providing a holistic snapshot of the key points (Wong, 1999). The different modes also provide appropriate entry points for students with different learning preferences, such as kinaesthetic, visual, or verbal mode of processing.

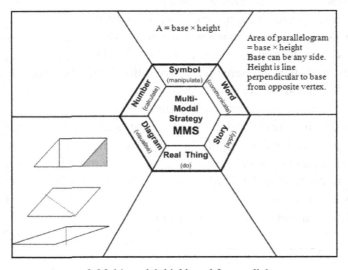

Figure 3. Multi-modal thinkboard for parallelogram

3.2 C = Confidence

When students have some success in mastering meaningful mathematics, their confidence will increase, and this will motivate them to strive for greater success in their study. Recent brain-based studies find that the endorphins released by the brain when students feel confident can increase neurological connections in the brain, leading to better learning.

It is not necessary that students must be immediately successful in every learning task, since learning from mistakes is also an important aspect of deep learning. The crucial point is that the students must not be stuck for too long without any feedback, mostly from the teacher; they need to see evidence that they are making some progress in the right direction. A personal anecdote may illustrate this point. I am not particularly motivated to play Sudoku because there is no intermediate feedback about whether the cells already completed are right or wrong, and it takes some time to finish one problem. However, on Web Sudoku[3], one can check the intermediate answers, and this immediately increases my motivation to play Sudoku. In a similar way, students are likely to be motivated when feedback is given along the way to give them the confidence to persevere to the end.

The cognitive feedback mentioned above works differently from behaviourist feedback based on praise and punishment. The behaviourist approach depends on extrinsic motivation such as praising students for good work, and this may raise their confidence in the short-term. On the other hand, punishing students for poor work may cause anxiety, loss of confidence, and avoidance of further engagement with mathematics.

The use of praise has generated much controversy in the literature. Some educators such as Hattie and Timperley (2007) and Kohn (1993) question its efficacy, while others feel that it has positive, small impacts in practice, though the effects may not be easily sustained in the long run. Also, the giving of praise and delivery of punishment may be culturally biased.

A different and widely cited approach of using praise is based on the research by the psychologist Carol Dweck (2006). She combines behaviourism and neuroscience to help students change their mindset about learning. Students who are praised for their *ability* (e.g., you are very smart) tend to develop a fixed mindset and fear of failure, and they tend to avoid challenging tasks. On the other hand, students who are

---

[3] http://view.websudoku.com

praised for their *effort* (e.g., you have put in a lot of hard work) tend to develop a growth mindset and are willing to try hard problems, thus becoming more actively engaged. Students can check their mindset at the Dweck Mindset website[4], and teachers can use this information to help their students develop the growth mindset. Her research suggests that simple "growth" messages such as "I believe you will work hard" may work. Perhaps, these growth messages can be supplemented by reflection questions, such as, "Which part of the problem did you work hard on and was your effort successful and why?"

Making mistakes is inevitable at the initial stage of mastering complex skills. However, students' belief about mathematical mistakes seems to be different from perceptions about mistakes in other complex skills such as sports and cooking. In the latter cases, the mistakes are rarely if ever marked with "crosses", and sometimes these mistakes as in the case of cooking can lead to pleasantly surprising outcomes like new flavours. Unfortunately, this does not work with mathematical mistakes, which are often marked wrong and given no credit, so the dominant student perception is to avoid mistakes at all cost. Furthermore, a behaviourist teacher may correct all the mistakes made by the students, thus highlighting failures rather than successes. Under this condition, many students readily give up on mathematics. Chinn (2012) noted that, all over the world, this seems to happen when students are about 7 years old! Those who experience more failures than successes are likely to be "motivated" to avoid taking risks to complete the tasks in order not to undermine their self-esteem. Chinn (2012) suggested that teachers pay attention to "no attempts" in student work as a sign of this behaviour. Thus, instead of correcting all the mistakes, a constructivist teacher, as suggested by von Glasersfeld (1998), may ask the students, "How did you get this answer?" to encourage them to decide for themselves whether the solutions are right or not; this "is the beginning of self-regulation ... the start of a potentially active learning process" (p. 28). It might be hard to "praise mistakes" as recommended by Posamentier and Jaye (2006), yet, teachers need to handle student mistakes with care. One way is to let them work on graded problems so that small success with

---

[4] http://mindsetonline.com/testyourmindset/step1.php

simple problems can stimulate students to tackle the more complex ones. Another way is to build a supportive environment that encourages students to take appropriate risks during learning.

The Trends in International Mathematics and Science Study, TIMSS 2011 (Mullis et al., 2012) measured students' confidence in mathematics using a 7-item scale with items such as "I usually do well in mathematics" and "I am good at working out difficult mathematics problems." The average scores of Singapore grades 4 and 8 students were 9.2 and 10.0 respectively, in the "somewhat confident" category. There is justification for raising confidence among Singapore students as a motivation strategy, but it is not clear which level of confidence is optimal to achieve outstanding performance. Nevertheless, it is prudent to take note of Bruner's (1960) warning: "The objective of education is not the production of self-confident fools" (p. 65).

3.3 *R = Relevance*

Relevance is a motivator that attempts to answer application questions from the students, such as "How to use this in my life?" and "Why study this?". When students think that the topics they are studying are important to their lives now (other than getting a good grade) as well as for the future, they might become interested in the assigned work. Two types of relevance can be discerned: cognitive relevance and emotion relevance.

Cognitive relevance
This focuses on the mathematics, which is what most educators think about when they argue for the relevance of mathematics. This includes applying mathematics to familiar contexts such as discount and currency exchanges, and getting students to do maths trails around their school and living environments is used for this purpose. It is also important to use unfamiliar real-life contexts so that students gain new knowledge about these contexts while at the same time honing their mathematical skills. Three recent local publications provide teachers with useful resources in this area: Ministry of Education (2012b); Ng and Lee (2012)

and Wong et al. (2012). Overseas examples (e.g., Maasβ, & O'Donoghue, 2011; Nagel, 1996; many examples from CSIRO maths-by-email[5]) can be adapted for local use.

The Singapore Mathematics Assessment and Pedagogy Project (SMAPP) includes two types of problems with real-life contexts, using authentic data as far as possible: (a) extended tasks with several parts covering skills, explanations, and reasoning, and (b) traditional problems focussing on well-defined skills (Wong et al., 2012). The contexts cover paper recycling, off-peak cars (a special scheme that Singapore implemented in 1994 to deal with car ownership and car use), water usage, mobile plans, tourism, and others. As an example, the decibel problem below was found by 39% of about 870 Secondary 1 Express (Grade 7) students to be relevant to their life. The context refers to the danger of hearing loss as a result of listening to loud music[6]. The question is as follows:

> The loudness of sound is measured in decibels (dB). Noise from heavy traffic is about 85 dB and this can cause hearing damage if one is exposed to it for 8 hours or more. For every 3 dB over 85 dB, the exposure time before damage occurs is decreased by half.
> (a) If the noise is 88 dB, what is the exposure time before damage occurs?
> (b) John likes to listen to his music using ear-plugs at high volume of 100 dB. How long could he do this before damage occurs?

Although this question does not include any real-life data, the context is authentic and of immediate concern to students, who will gain new knowledge that might help them make appropriate decisions when they listen to music. For this student sample, only 27% were able to solve (a) with correct working and answer, while another 10% obtained the correct answer without showing working. Part (b) was more difficult with only 21% getting full credit, and 45% gave the wrong answer. The popular method was stepwise decrease of the number of hours with each

---

[5] http://www.csiro.au/resources/Maths-by-Email.html

[6] For example, see http://www.noisehelp.com/noise-dose.html

3 dB increment. For students at higher grade levels, the definition of decibel may be given (an increase of 10 dB in loudness, *L*, corresponds to a 10-fold increase in sound intensity, *I*), and they can be asked to express *I* in terms of *L*, thus illustrating an authentic application of the exponential and logarithmic functions.

Cognitive relevance is also about developing a mathematical lens to query quantitative information or statistical inferences reported in the mass media. One example from SMAPP is about the salt content of a local dish given in a newspaper report: 1,675 g. About two thirds of the Secondary 1 Express students did not spot this mistake. This shows that many students have not developed an intuitive sense about the magnitudes of common quantities, despite having solved numerous arithmetic and mensuration problems, because they do not ask themselves whether the answers they have computed make sense or not. This lack of number sense about various types of quantities is also found in adults including professionals. The author has tested the following question (based on Gigerenzer, 2002) on in-service mathematics teachers and most of them could not give a reasonable initial guess of the answer. Some of them tried to use Bayes' Theorem with limited success as they could not recall the formula.

The probability of a certain disease in the population was 0.005. The probability of a positive test on someone with the disease was 0.95 and the probability of a negative test on someone without the disease was 0.92. What is the probability that someone has the disease given a positive test?

It is easy to set up an Excel worksheet to explore the answers by changing the given information. This will foster a stronger sense of risks in real life situations, and this new knowledge will help people develop a better appreciation of the benefits of various types of health screening. Students are likely to be motivated to explore similar real-life situations under suitable guidance.

TIMSS 2011 (Mullis, et al., 2012) had a 6-item Students Value Mathematics scale, with items such as "I think learning mathematics will help me in my daily life" and "I need to do well in mathematics to get the job I want." Only grade 8 students responded to this scale. The average score of the Singapore sample was 10.0, in the "somewhat value" category. The 43% who valued mathematics had significantly higher mathematics scores (619) than the 10% who did not value mathematics (591). Although appreciation of the values of mathematics does not necessarily cause higher mathematics achievement, the relationship is strong enough to recommend including this value as an important motivator.

Wong (2008) argued that knowledge about real-life contexts is necessary to solve certain types of application and modelling problems, and this should be included as an important factor in mathematics problem solving. This approach is emphasised in the extended curriculum framework shown in Figure 4. This framework also stresses the aim of helping students gain new knowledge about the world at the same time when they solve mathematics problems.

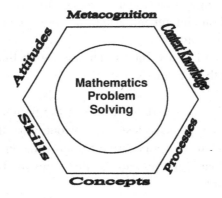

Figure 4. An extended mathematics curriculum framework for Singapore

Emotion relevance

This refers to the connection of learning experiences to emotions. According to neuroscience, incoming stimuli may pass through the emotion region of the amygdala in the brain before they reach the prefrontal cortex for cognitive processing and decision making. Hence, "[e]motions are the gatekeepers to the intellect" (Pete & Fogarty, 2003, p. 13), and with "emotions affecting rational decision-making" (Evans,

2001, p. 133), emotion relevance may be more critical than cognitive relevance as a significant motivator. Tasks that have emotion relevance are often enjoyable and fun to the students, and completing these tasks can satisfy their curiosity about the world as well as mathematics (though to a less extent in the latter case). The next section addresses this motivator in greater detail.

3.4 *E = Enjoyment*

Participating in enjoyable activities is a powerful motivator, especially when enjoyment also satisfies curiosity. Young children are naturally curious about many things and they try to satisfy their curiosity by asking numerous "why" questions. But as noted by the early childhood educator Lillian Weber (cited in Brown, 2008, p. 63), students "begin school an exclamation point and a question mark; often they leave as a plain period." Unfortunately, many students seem to lose their curiosity as they advance through the school system.

This also happens to attitudes toward learning mathematics. Research has shown "the tendency for attitude scores [toward mathematics] to decline as students move from elementary school to secondary school" (McLeod, 1994, p. 639). This trend was found in many countries in TIMSS 2011, as measured using a 5-item scale, with items such as "I enjoy learning mathematics" and "I learn many interesting things in mathematics." The authors of TIMSS noted: "Compared to the fourth grade, substantially fewer eighth grade students reported positive attitudes toward learning mathematics" (Mullis et al., 2012, p. 328). Quite surprisingly, this was not the case with Singapore students: the average scores of Singapore 4th and 8th grade students were 9.9 and 10.4 respectively, in the "somewhat like learning mathematics" category. Many reasons may be offered to explain this deviation from the international pattern. However, the important implication for teachers is how to provide the 20% of Singapore students who may not like mathematics with enriched experiences to enhance their attitudes. Several examples are discussed below.

Curiosity is the urge to find reasons for unexpected situations, and mathematics is actually full of counter-intuitive and surprising results. For examples, why is the ratio of the circumference of a circle to its diameter a constant, irrespective of its size? Why is zero related to one as in $a^0 = 1$? Why does Pythagoras Theorem work using squares of sides but not cubes of sides? The surprises these results might generate seem to be lost when teachers present them as "standard" results to be learned by heart and used to solve routine problems. To avoid this loss of unexpectedness, teachers may make the mathematics results more meaningful and act out the surprises they inspire. As an example, the difference of two squares may be given as just another algebraic rule. Instead, present students with the following diagram (Figure 5) and note that: $6 \times 8 = 48$, one less than $7^2 = 49$. Does this pattern work with any three consecutive integers? What happens when the three numbers differ by two instead of one, say, 6, 8, and 10? Hopefully, this is puzzling enough to make the learning more enjoyable than just follow given rules.

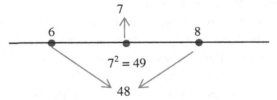

Figure 5. Difference of two squares

The above activity may be followed by asking students to think about what Fibonacci wrote more than 700 years ago: a square number exceeds its preceding square number by the sum of the two roots. As a child living in Algeria, Fibonacci learned the Hindu-Arabic number system, which he later introduced to Europe. He is now best remembered for solving the rabbit reproduction problem, resulting in the sequence {1, 1, 2, 3, 5, ...} named after him. Adding interesting historical anecdotes is another way to make lessons more appealing to students.

Many students find proving and applying Euclidean geometry properties abstract and tedious. This learning experience can be made more enjoyable by helping students make sense of the geometry properties through hands-on activities such as measuring, cutting, and folding (the *M* motivator), followed by tinkering with the results. To illustrate, consider the Angle Bisector Theorem: in Figure 6, if *AD* bisects angle *A*, then, *BD : DC = AB : AC*.

First, let students draw a few triangles, construct the angle bisectors, and measure the sides as accurately as possible. Check that the result is correct within measurement errors. Next, re-arrange the ratio to link to previous knowledge. Re-writing the ratio as $\frac{BD}{AB} = \frac{DC}{AC}$, it "looks" like sin θ. One can also "cross multiply" the sides as *AB* × *DC* = *AC* × *BD*. Such tinkering helps students to become more familiar with the given result and remember it better, thus making learning both more meaningful and enjoyable.

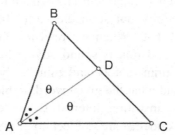

Figure 6. Angle bisector theorem

Thinking about unusual events can be intriguing and fun. Textbook examples and real-life situations about discount inevitably involve percentages less than 100%. Pose this question: Does it make sense to have discounts of 100% or more? Several years ago, in an effort to attract more customers, a restaurant in Singapore advertised an age-based discount, namely, discounts were given on food bills equal to the age of the paying customer. A great grandmother 104 years old celebrated her birthday with her family at the restaurant. How would the restaurant handle a discount of 104%? This was reported in the *Straits Times* (16 May 2009). Teachers can collect similar unusual examples and use them to enliven their mathematics lessons.

Other ways to make lessons more enjoyable include: sing songs about mathematics to aid memory[7]; share interesting stories about mathematicians from the West and East (Ramsden, 2009; Reimer & Reimer, 1995); use mathematics humour and cartoons (e.g., Paulos, 1980; Potter, 2006; Reeves, 2007; Wong, 2001) in lessons and link mathematics to arts, music, and stamps (Tan, Wong, & D'Rozario, 2002), such as Escher designs, Mandelbrot patterns, and tessellations. Mathematics stories should include problems that students can try to solve. Three well-known examples are: young Gauss, teacher, and the sum of first 100 natural numbers; young and sickly Descartes, housefly and the coordinate system; Ramanujan, taxi, and 1729. Mathematicians are also known for their compassion and wisdom. For example, Pascal took in a homeless family but soon died after being infected by that family. He also noted that "When we read too fast or too slowly, we understand nothing"[8], and this is sound advice to share with students, especially those who immediately and mindlessly apply rules without first trying to understand what the given word problems are about.

An impediment to instilling interest is the enormous amount of routine practice that students are subject to. Extensive and deliberate practice is critical for the acquisition of skills, but it must not be so demanding that it prematurely "kills" motivation via the route of enjoyment.

Interest and confidence are often interrelated. It is helpful to recall the advice that Piaget (1973) gave several decades ago:

> every normal student is capable of good mathematical reasoning if attention is directed to activities of his [sic] *interest* [italics added], and if by this method the emotional inhibitions that too often give him [sic] a feeling of *inferiority* [italics added] in lessons in this area are removed. (pp. 98 – 99)

---

[7] For a song about quadratic equation sung by Singapore students, see http://www.youtube.com/watch?v=VS56N94Sllc

[8] http://en.wikiquote.org/wiki/Blaise_Pascal

According to Marzano (2013), "[t]teacher can also indirectly communicate the importance of content through their enthusiasm ... by recounting how they became interested in the content when they were students themselves" (p. 82). Teacher enthusiasm was found to mediate the positive link between teacher enjoyment in teaching and student enjoyment in learning within grades 7 and 8 mathematics classrooms in Germany (Frenzel et al., 2009). Generating mutual interest in mathematics tasks can strengthen rapport between teachers and students, which is the fifth motivator of *M_Crest*, to be discussed in the next section.

3.5 *S* = *Social relationships*

As social beings, children and adults have strong senses of belonging to specific communities, and interacting with others can enhance personal well-being. This powerful motivator can occur in three different ways in the schools.

First, there should be mutual rapport between the teacher and the students. In fact, many students are motivated to study because they like to please teachers who care about them as individuals rather than just about whether they have learned the mathematics. The Nobel laureate in physics Richard Feynman (1963/2005) believed that "the best teaching that can be done only when there is a direct individual relationship between a student and a good teacher" (p. xxix). This concurs with the Teachers' Vision set by the Singapore Ministry of Education in 2009: *We lead, care, inspire.* This rapport will provide an environment conducive to meaningful learning.

Second, many students like to work in groups for various purposes, such as to help one another learn, to develop social skills, to make friends, to gain recognition among peers. When in doubt, many students prefer to ask their classmates rather than the teacher, thus providing further reason why strengthening social relationships among students is an influential motivator. Teachers can tap into peer influences by encouraging *groups*, not just individuals, to like mathematics. However, Cain (2012) noted that a sizeable proportion of students are introverts, and these students may prefer to complete tasks on their own rather than work in groups. She recommended rightly that teachers use a balance of group and independent activities to serve all types of students in their classes.

Third, nowadays many students are IT savvy and they use social networking tools with considerable ease. The Ministry of Education (2012a) recommends that students share reflections about their learning through blogs because doing so "makes learning social" (p. 25). Engaging learning through this social media can be motivating for the digital natives.

3.6 *T = Targets (goals)*

This is the last motivator in the *M_Crest* framework, and it focuses on helping students develop and achieve learning targets that they could and should aspire to. There are immediate, short-term, and long-term targets or goals.

Immediate targets are basically about completing assigned tasks, such as seatwork and homework, within reasonable time limits, which are usually motivating (Steele, 2009). Students may complete these tasks to please the teachers or parents. Doing so is an extrinsic motivator that works for students who understand the cultural mores about attending schools. They aim to gain praises and rewards and to avoid punishments. This gives rise to the "token economy" in numerous classrooms. Under this system, teachers give tangible rewards in exchange for completed tasks and desirable behaviours. Monetary rewards are commonly used but their longer-term effects are questionable. Fryer (2010) reported a recent study involving about 15,000 grades 4 and 7 students in New York. These students were paid money for good performance, for examples, $10 for completing a test and $50 for getting a perfect score. After a year of this expensive intervention (US $6 million), the average earning was $231 per student. Yet, there were no significant differences between these students and those in control classes in test scores, effort to study, and intrinsic motivation. Indeed, similar behaviours are observed among adults: the political philosopher Michael Sandel (2012) has shown by many examples, how placing a price on things may be efficient but it also damages the inculcation of desirable norms and attitudes. This sends a strong message against setting monetary targets as a means to promote desirable changes among students and even teachers.

Short-term targets take months to achieve. These include mastering the curriculum contents and processes at the grade level. Students often do not explicitly think about these objectives, preferring to focus on the tasks on hand. It is worthwhile to help them understand the importance of these short-term goals so that they do not lose heart over present failures as these can be taken as learning opportunities over a longer period. Thus, they should focus on mastery goals, namely to do one's best in mastering the tasks, rather than performance goals, such as getting a good grade in the upcoming test or avoiding making mistakes at all cost.

Long-term targets refer to those in the future such as career aspirations. These are often stated as broad aims in subject syllabuses. The Singapore mathematics curriculum (Ministry of Education, 2012a) states that the broad aims of mathematics education over the 12 years of schooling are to enable students to:

- acquire and apply mathematical concepts and skills;
- develop cognitive and metacognitive skills through a mathematical approach to problem solving; and
- develop positive attitudes towards mathematics. (p. 7)

Both cognitive and affect goals are included. These goals cannot be accomplished within a few lessons. They may be realised, if at all, only after several years of studying mathematics. Teachers must not lose sight of these long-term aims and should try their best to remind students of these targets as the students move from one grade to another. Such reminders may provide strong motivation over many years and may even arrest the deterioration in attitudes toward learning mathematics with grade levels, as discussed in earlier section.

The idea of autonomy from self-determination theory (Ryan & Deci, 2000) suggests that in a democratic society, motivation can be enhanced when students are given some degree of autonomy to determine for themselves the targets they wish to achieve, the ways to achieve them, when to do so, with whom they work with, and even judging the quality of their work. This strong sense of choice is not very practical in an examination oriented system like Singapore. Nevertheless, teachers may be able to motivate students by giving them small choices some of the time, for example, which problems in a given set of parallel problems to do. This approach helps students develop some degree of self-regulation of their learning, which is one of the two aspects of metacognition in the Singapore mathematics framework.

3.7 *A Combination of Motivators*

The six motivators in *M_Crest* become truly powerful when they are applied together rather than separately. Many different combinations of these motivators are possible, depending on students' past experiences, present needs, and the pedagogical skills of the teachers. As an example of a feasible combination, students are likely to enjoy (*E*) their lessons when they can make sense (*M*) of and understand the relevance (*R*) of what is to be learned, set within at least some short-term goals (*T*). Their motivation can be further enhanced if they are allowed to work together (*S*), if so desired, under the guidance of a caring teacher (*S*). Successful learning will build up their confidence (*C*) to tackle further learning.

Considerable pedagogical expertise is required to decide which combination of motivators to use, when to use to them, and for whom. Reading about and listening to the experiences of others can be useful in terms of knowledge gained, but this kind of knowledge does not necessarily lead to real benefits for students unless the teachers experiment with different combinations of the motivators, master the techniques with frequent practice, reflect on what works or does not work in their lessons, and make the necessary refinements. A systematic way to do so is to engage in action research.

4 Teacher Research into Motivation to Learn

Action research is widely promoted as a powerful form of teacher professional development (e.g., Baumfield, Hall, & Wall, 2008; Masingila, 2006). In Singapore, action research was introduced in the

1980s, became dormant in the 1990s, and was again promoted in the 2000s with strong support from the Ministry of Education. Once the teachers have acquired the essential *research* skills, they can then design *action* plans to test the efficacy of various motivating strategies.

The earlier sections have mentioned three scales used by TIMSS to assess confidence, values, and enjoyment of mathematics. To use *M_Crest* as a framework for motivation research, it is necessary to develop scales to measure the remaining three motivators. Sample items are given below:

- It is important that Mathematics makes sense to me. [*M*]
- I am more engaged when I work with my friends. [*S*]
- I prefer to work from simple to difficult problems. [*T*]

A typical research question concerns the relationships between motivation and achievement. For example, Lim (2010) investigated this relationship among 984 Junior College students (grades 11 and 12) and found that the correlations with mathematics scores were highest with self-confidence (0.60) and lowest with extrinsic motivation (-0.05). Teachers may replicate this study to understand how these relationships work for their own students. However, these relationships are not causal in nature. More importantly, experiments should be conducted to investigate the effects of specific motivators on student learning, for example, using a pre-post-test design where both quantitative and qualitative data are collected.

Another approach is for teachers to reflect on their own practices on a weekly basis by writing down their thoughts in a journal. Some reflection questions are:

- What feedback did I seek from students about their motivation?
- What had I done to motivate (whole class, groups, individual students) using the (rewards, ...) motivators?
- How might this reflection change my plan for the following week to engage the reluctant or disengaged students?

Sharing their reflections with colleagues within a professional learning community is also very beneficial. For examples, teachers who teach the same students for different subjects may get together to compare what motivators work or do not work for these students. Through this sharing, the teachers will gain deeper understanding of the different needs and preferences of the same students under different learning contexts and expectations so that they can plan more effective lessons to motivate them to learn.

The following proposal is inspired by a study reported by Amabile in 1985 about writing and cited by Ramsey and Schafer (2013). The hypothesis to be tested is based on the psychological theory of "priming" and examines whether inducing students to think about intrinsic or extrinsic motivation affects their mathematical scores. Assign students randomly to two groups, one taking an intrinsic questionnaire and the other an extrinsic questionnaire. Each questionnaire has about 5 items, such as:

- (Intrinsic) I enjoy solving mathematics problems.
- (Intrinsic) I like to play with numbers.
- (Intrinsic) I gain new skills by solving mathematics problems.
- (Extrinsic) My parents encourage me to do my best to solve mathematics problems.
- (Extrinsic) I always have good grades in mathematics.
- (Extrinsic) My teacher is impressed with my ability to solve difficult problems.

The students rank these reasons in the respective questionnaire. After completing the questionnaire, they solve the same set of mathematics problems. The mathematics scores of the two groups can be compared to test the hypothesis. The outcomes could stimulate further investigation into the priming effects of different motivators on problem solving, which is an uncharted research area.

A common belief is that motivation leads to actions and engagement in learning tasks. However, the reverse also happens. Students may begin to tackle a learning task grudgingly, but as they get more involved and gain familiarity with it, their motivation may be ignited. This results in "actions lead to motivation". This hypothesis is worthy of inquiry.

5 Conclusion

The six motivators included in the *M_Crest* framework provide powerful *reasons* to *move* students to *study* mathematics. According to Dictionary.com[9], one etymological root for the word *study* is the Latin *studium,* which means "eagerness". Students must see *meaning, relevance, enjoyment,* and clear *targets* in what they are studying in order to generate the eagerness or motivation to undertake the study. Another root is the Latin *studere,* meaning "to be diligent or to press forward". *Confidence* and positive *social relationships* can help students sustain the effort and perseverance to press forward to tackle challenges encountered in their learning journey. Eventually a successful student is one who is self-motivated. This is what every caring teacher should strive to foster in all the students he or she has the privilege to work with.

References

Ausubel, D. P. (1968). *Educational psychology: A cognitive view*. New York, NY: Holt, Rinehart and Winston.

Baumfield, V., Hall, E., & Wall, K. (2008). *Action research in the classroom*. London: Sage Publications.

Brahier, D. J., & Speer, W. R. (Eds.). (2011). *Motivation and disposition: Pathways to learning mathematics*. Reston, VA: National Council of Teachers of Mathematics.

Brown, S. (2008). *A Buddhist in the classroom*. Albany, NY: State University of New York Press.

Bruner, J. S. (1960). *The process of education*. Cambridge, MA: Harvard University Press.

Cain, S. (2012). *Quiet: The power of introverts in a world that can't stop talking*. New York, NY: Crown Publishers.

---

[9] http://dictionary.reference.com/browse/study

Caine, R. N., Caine, G., McClintic, C., & Klimek, K. (2009). *12 brain/mind learning principles in action: The fieldbook for making connections, teaching, and the human brain* (2nd ed.). Thousand Oaks, CA: Corwin Press.

Chinn, S. (2012). *The trouble with maths: A practical guide to helping learners with numeracy difficulties* (2nd ed.). London: Routledge.

de Jong, T., van Gog, T., Jenks, K., Manlove, S., van Hell, J., Jolles, J., van Merrienboer, J., van Leeuwen, T., & Boschloo, A. (2009). *Explorations in learning and the brain: On the potential of cognitive neuroscience for educational science.* Dordrecht: Springer.

Dienes, Z. P. (1964). *Building up mathematics* (2nd ed.). London: Hutchinson Educational.

Dweck, C. S. (2006). *Mindset: The new psychology of success.* New York, NY: Random House.

Evans, D. (2001). *Emotion: The science of sentiment.* Oxford: Oxford University Press.

Feynman, R. P. (2005). *Six easy pieces: Essentials of physics explained by its most brilliant teacher.* New York, NY: Basic Books. (Original work published 1963)

Frenzel, A. C., Goetz, T., Lüdtke, O., Pekrun, R., & Sutton, R. E. (2009). Emotional transmission in the classroom: Exploring the relationship between teacher and student enjoyment. *Journal of Educational Psychology, 101*(3), 705-716.

Fryer, R. G., Jr. (2010). *Financial incentives and student achievement: Evidence from randomized trials.* Retrieved from http://scholar.harvard.edu/files/fryer/files/financial_incentives_and_student_achieve ment_evidence_from_randomized_trials.pdf

Gigerenzer, G. (2002). *Reckoning with risk: Learning to live with uncertainty.* London: Penguin Books.

Hattie, J. A. C. (2009). *Visible learning: A synthesis of over 800 meta-analyses relating to achievement.* London: Routledge.

Hattie, J., & Timperley, H. (2007). The power of feedback. *Review of Educational Research, 77*(1), 81-112.

Kohn, A. (1993). *Punished by rewards: The trouble with gold stars, incentive plans, A's, praise, and other bribes.* Boston, MA: Houghton Mifflin.

Lim, S. Y. (2010). Mathematics attitudes and achievement of Junior College students in Singapore. In L. Sparrow, B. Kissane, & C. Hurst (Eds.), *Shaping the future of mathematics education: Proceedings of the 33rd annual conference of the Mathematics Education Research Group of Australasia* (pp. 681-689). Fremantle: MERGA.

Maasβ, J., & O'Donoghue, J. (Eds.). (2011). *Real-world problems for secondary school mathematics students: Case studies.* Rotterdam, the Netherlands: Sense Publishers.

Marzano, R. J. (2013). Ask yourself: Are students engaged? *Educational Leadership, 70*(6), 81-82.

Masingila, J. O. (Ed.). (2006). *Teachers engaged in research: Inquiry into mathematics classrooms in Grades 6-8.* Greenwich, CN: Information Age Publishing.

McLeod, D. B. (1994). Research on affect and mathematics learning in the JRME: 1970 to the present. *Journal for Research in Mathematics Education, 25*(6), 637-647.

Middleton, J. A., & Jansen, A. (2011). *Motivation matters and interest counts: Fostering engagement in mathematics.* Reston, VA: National Council of Teachers of Mathematics.

Ministry of Education, Singapore. (2012a). *O-level mathematics: Teaching and learning syllabus.* Singapore: Author.

Ministry of Education, Singapore. (2012b). *Problems in real-world contexts: A resource book.* Singapore: Author.

Mullis, I. V. S., Martin, M. O., Foy, P., & Arora, A. (2012). *TIMSS 2011 International results in mathematics.* Chestnut Hill, MA & Amsterdam, the Netherlands: TIMSS & PIRLS International Study Center, Lynch School of Education, Boston College & International Association for the Evaluation of Educational Achievement.

Nagel, N. G. (1996). *Learning through real-world problem solving: The power of integrative teaching.* Thousand Oaks, CA: Corwin Press.

Ng, K. E. D., & Lee, N. H. (Eds.). (2012). *Mathematical modelling: A collection of classroom tasks.* Singapore: Alston Publishing.

Paulos, J. A. (1980). *Mathematics and humor.* Chicago: University of Chicago Press.

Pete, B. M., & Fogarty, R. J. (2003). *Twelve brain principles that make a difference.* Thousand Oaks, CA: Corwin Press.

Piaget, J. (1973). *To understand is to invent: The future of education.* (G.A. Roberts, trans.). New York, NY: Grossman Publishers.

Pink, D. H. (2009). *Drive: The surprising truth about what motivates us.* New York, NY: Riverhead Books.

Posamentier, A. S., & Jaye, D. (2006). *What successful math teachers do, grades 6-12: 79 research-based strategies for the standards-based classroom.* Thousand Oaks, CA: Corwin Press.

Posamentier, A. S., & Krulik, S. (2012). *The art of motivating students for mathematics instruction.* New York, NY: McGraw Hill.

Potter, L. (2006). *Mathematics minus fear.* London: Marion Boyars.

Ramsden, J. (2009). *Whoever thought of that?: Short biographies of some contributors to the history of mathematics.* Leicester: Mathematical Association.

Ramsey, F. L., & Schafer, D. W. (2013). *The statistical sleuth: A course in methods of data analysis* (3rd ed.). Australia: Brooks/Cole Cengage Learning.

Reeves, A. (Ed.). (2007). *Cartoon corner: Humor-based mathematics activities.* Reston, VA: National Council of Teachers of Mathematics.

Reimer, L., & Reimer, W. (1995). *Mathematicians are people, too: Stories from the lives of great mathematicians: Vol. 2.* Palo, Alto, CA: Dale Seymour.

Russell, B. (1975). *The autobiography of Bertrand Russell.* London: Unwin Paperbacks.

Ryan, R. M., & Deci, E. L. (2000). Intrinsic and extrinsic motivations: Classic definitions and new directions. *Contemporary Educational Psychology, 25,* 54-67.

Sandel, M. (2012). *What money can't buy: The moral limits of markets.* London: Allen Lane.

Sansone, C., & Harackiewicz, J. M. (Eds.). (2000). *Intrinsic and extrinsic motivation: The search for optimal motivation and performance.* San Diego, CA: Academic Press.

Schunk, D. H., Pintrich, P. R., & Meece, J. L. (2008). *Motivation in education: Theory, research, and applications* (3rd ed.). Upper Saddle River, NJ: Pearson/Merrill Prentice Hall.

Sousa, D. A. (2008). *How the brain learns mathematics.* Thousand Oaks, CA: Corwin Press.

Steele, C. F. (2009). *The inspired teacher: How to know one, grow one, or be one.* Alexandria, VA: ASCD.

Stipek, D. J. (2002). *Motivation to learn: Integrating theory and practice* (4th ed.). Boston, MA: Allyn and Bacon.

Tan, W. K., Wong, A. F. L., & D'Rozario, V. (2002). *Philatelic forays across the curriculum.* Singapore: Singapore Philatelic Museum.

von Glasersfeld, E. (1998). Why constructivism must be radical. In M. Larochelle, N. Bednarz, & J. Garrison (Eds.), *Constructivism and education* (pp. 23-28). Cambridge: Cambridge University Press.

Wong, K. Y. (1999). Multi-modal approach of teaching mathematics in a technological age. In E. B. Ogena & E. F. Golia (Eds.), *8th Southeast Asian Conference on Mathematics Education, technical papers: Mathematics for the 21st century* (pp. 353-365). Manila: Ateneo de Manila University.

Wong, K. Y. (2001). Mathematics cartoons and mathematics attitudes. *Studies in Education, 6,* 69 – 80.

Wong, K. Y. (2008). An extended Singapore mathematics curriculum framework. *Maths Buzz, 9*(1), 2-3.

Wong, K. Y., Zhao, D. S., Cheang, W. K., Teo, K. M., Lee, P. Y., Yen, Y. P., Fan, L. H., Teo, B. C., Quek, K. S., & So, H. J. (2012). *Real-life mathematics tasks: A Singapore experience.* Singapore: Centre for Research in Pedagogy and Practice, National Institute of Education, Nanyang Technological University.

Woolfolk, A., Hughes, M., & Walkup, V. (2013). *Psychology in education* (2nd ed.). Harlow: Pearson Education.

Chapter 3

Designing Learning Experiences for Effective Instruction in Secondary Mathematics

TOH Tin Lam

In the recent review of the Singapore secondary mathematics curriculum, *learning experience* has been incorporated as one of the main emphases besides the revision in the mathematical content. Learning experiences are explicitly stated in the mathematics syllabus documents to influence the ways teachers teach and students learn in order to achieve the curriculum objective. This chapter discusses (1) the principles of selecting appropriate learning experiences and their characteristics; and (2) how these learning experiences can be organized for effective instruction in the secondary school mathematics classrooms. The model of Tyler (1946) will be the framework for discussion in this chapter. The discussion will use illustrations from the Singapore lower secondary mathematics classrooms.

1 Introduction

In the recent Singapore secondary mathematics curriculum revision, *learning experience* has been identified as one of the new emphases in the mathematics curriculum. It is a message to mathematics teachers that *how* students learn mathematics is at least as important as *what* students learn in the curriculum, if not more important. It is the aim of the Singapore mathematics curriculum that "[t]hese processes [i.e. the cognitive and the metacognitive skills] are learned through carefully constructed learning experiences." (Ministry of Education, 2012).

Although it has always been recognized by teachers, educators and researchers in Singapore that both processes and product are equally important in education, the emphasis on the process of learning is now committed in print in the official syllabus document. These statements about learning experiences, which are now expressed in the form of "students should have opportunities to…", serve to remind teachers of the student-centric nature of these experiences (Ministry of Education, 2012). Examples of such learning experiences associated with each topic in the syllabus are described clearly in the new syllabus document.

The examples given in the syllabus documents are meant for teachers to implement in mathematics classrooms. These few examples are not sufficient. Teachers must be able to design appropriate learning activities to enhance their students' learning. This chapter suggests the principles of selecting appropriate learning experience and organization of learning experience. These are two important things teachers must know, based on Tyler's model.

2 Learning Experience

In the ethical philosophy of Aristotle, "the importance of being experienced" plays the pivotal role in education. The educative experiences are valuable in terms of both having a forming power on the subject and widening up a person's action possibilities (Saugstad, 2013). Although Aristotle's understanding of educative experience, being rooted in a culture-based understanding, needs further interpretation today, educators would generally agree with the importance of experience in education for an individual.

In the early part of the twentieth century, the concept of *learning experience* (or *learning activities*) was completely absent from the education literature. With the evolution of the field of psychology, learners' activities received attention from educators. Subsequently, the term *learning activities* was replaced by *learning experience*, as the former could not adequately describe the dynamics of the entire teaching-learning situation. For example, two students undergoing the same *learning activity* might not have the same learning *experience*.

Taba (1962), one of the well-known figures in curriculum development, argued that there was a definite order to creating the curriculum. She proposed that the selection of suitable learning experiences is one of the key steps in creating a curriculum. According to Taba, students must engage with the content presented to them. One of the key roles of the teachers is to select and organize appropriate learning activities and experiences suitable for their students. If curriculum content is the "meat" of the curriculum, then according to Taba, learning experiences must be described as the "heart". In fact, Taba noted that

> perhaps the first important consideration in achieving a wider range of objectives is the fact that the learning experiences, and not the content as such, are the means for achieving all objectives besides those of knowledge and understanding (p. 278).

Taba's work can be seen as a further development of the pioneering work on curriculum development developed by Ralph Tyler (1946). Tyler is well-known for his approach in thinking about education and curriculum. His model of curriculum design and development is one of the best known among the technical-scientific models of curriculum design and development. In the third of the four basic principles in thinking about curriculum proposed by Tyler, it was stated that people involved in curriculum inquiry must try to define educational experiences related to the purposes of the curriculum. According to Tyler (1946), learning experience involves the interaction of the learner and his or her environment, resulting in some features of his environment attracting his or her attention and to which he or she reacts. Implied in this definition of learning experience is that the learner is an active participant in the process of learning.

In the field of mathematics education, it has been recognized by researchers and educators that good learning dispositions can be cultivated in students by establishing rich and rewarding mathematical experiences, and that ignoring such dispositions in the learner can be detrimental (for example, Barton, 1993).

In this chapter, Tyler's principles for selecting appropriate learning experiences and how these learning experiences can be organized for effective learning will be presented in the context of teaching secondary school mathematics. Specific examples from selected mathematics topics will be used for illustration.

3 Principles for Selecting Appropriate Learning Experiences

Tyler (1946) identified five principles of selecting appropriate learning experiences for achieving particular objectives in the curriculum. These five principles are:

- Principle One: for a particular objective to be attained, students must have experiences that give him an opportunity to practice the behaviour implied by the objective;
- Principle Two: students must obtain satisfaction from the selected learning experiences;
- Principle Three: responses expected from students for the learning experience must be within the means for the students;
- Principle Four: there are many experiences to achieve the same objective; and
- Principle Five: the same learning experience can bring about several outcomes.

In this section, lower secondary math topics will be used to illustrate how these principles can be incorporated into the planning of learning experiences for mathematics instructions.

3.1 *Principle One*

There are two key aspects for discussion with regards to Principle One.

Mathematical problem solving
The heart of the Singapore mathematics curriculum is problem solving.

As such, students must be provided with ample opportunities in the mathematics classrooms to practice problem solving. What is a 'problem'? A 'problem' has multiple and often contradictory meanings through the years. According to the generally accepted definition in the field of mathematics education, a problem is a situation in which an individual or group is called upon to perform a task for which there is *no readily accessible algorithm* which determines completely the method of solution. Lester (1980) adds that this definition of a problem must assume a desire on the part of the individual or group to perform the task.

Teachers have been phenomenally successful in routinizing the mathematics problems that students encounter in the present high-stake mathematics examination; this method has proved to be successful in view of the large number of distinctions over the past decades in the various national mathematics examinations. Based on this approach, students might be exposed to solving mathematics questions which have been routinized by their teachers. Although many of the school teachers may claim some support from the assertion by Siu (2001) that one of the purposes of mathematics is to make real world problems routine, this approach of making problems routine seems to be totally against the spirit of problem solving in the mathematics curriculum.

Many teachers might not have a complete understanding of problem solving. Some Singapore mathematics teachers might have reduced the teaching of problem solving in schools to the teaching of heuristics (Hedberg et al., 2005). The right choice and use of heuristics for any particular mathematics question were assumed to be sufficient for successful problem solving. If this is the belief held by most teachers, it would not be surprising that students would be required to spend most of their time identifying the correct heuristics to each category of mathematics problems. This also appears to be against the spirit of true problem solving.

Problem solving skill refers to the ability of an individual to get "unstuck" when he or she encounters a (nonroutine) problem. In line with the spirit of mathematical problem solving, students must be explicitly taught a problem solving model and given sufficient curriculum time to solve nonroutine mathematics problems by applying the problem solving model. Any student attempting mathematical problem solving requires a model to which he or she can refer, especially when progress is not satisfactory. Good problem solvers would presumably have built up their own models of problem solving. In equipping students with problem solving ability, it is first crucial to explicitly teach them with a model of problem solving (in the Singapore mathematics curriculum, Polya's model is recommended, although any other model of problem solving will be equally viable). A problem solving model that is made explicit to students should be helpful in guiding them in the learning of problem solving, and in regulating their problem solving attempts. Even a good problem solver may find the structured approach of a model useful (Schoenfeld, 1985).

To take problem solving to its truest spirit, it is essential for students to be given sufficient opportunity to be exposed to genuine problem solving by attempting to solve unseen problems via using a problem solving model. For a full learning experience on problem solving, through a module designed specifically to enable students to experience problem solving, readers can refer to Toh et al. (2011), which gives a full description of the activity involved in problem solving. Such problem solving lessons give students the opportunity to put the problem solving model into practice in tackling unseen problems. In short, the learning experience associated with such genuine problem solving should not be replaced by teachers' routinizing of problems into classroom exercises together with excessive drill-and-practice.

Mathematical literacy

Mathematical Literacy is attracting worldwide attention, following the development of the PISA 2012 mathematics framework. *Mathematical literacy* is defined as

> an individual's capacity to formulate, employ, and interpret mathematics in a variety of contexts. It includes reasoning mathematically and using mathematical concepts, procedures, facts, and tools to describe, explain, and predict phenomena. It assists individuals to recognise the role that mathematics plays in the world and to make the well-founded judgments and decisions

needed by constructive, engaged and reflective citizens (OECD, 2010, p. 4).

Teachers must provide adequate opportunities in the mathematics classrooms to allow students to be able to perform the tasks described in the above definition of mathematical literacy. This section presents several illustrations on how mathematical literacy can be developed in the mathematics classrooms by incorporating appropriate learning experience. In short, students should be challenged to work on the application of mathematics beyond mere computation. Two ways that this can be done are:
- "reading" mathematics; and
- making decisions in real-life situations.

Reading mathematics in real-world context
It has been teachers' experience that students want to know how mathematics is relevant to their daily lives. The call among the mathematics education community throughout the world to introduce mathematical tasks that are related to 'real life' and the 'real world' could be traced back to as early as 1982 in the Cockcroft Report about the increasing concern that adults were not able to apply the mathematics they had learned at school in everyday contexts (Boaler, 1993). One of the five main reasons provided by the advocates of this movement is that the relevance of mathematics to students' daily lives would enhance students' appreciation of mathematics; and thereby improve their affect in the subject (Beswick, 2011).

In classroom situation, students should be given ample opportunity to "read" mathematics in the real life context, use real-world data to provide discussion among students and interpret mathematical information in the real-world context. For example, consider the following comic strip, shown in Figure 1, which was used as a chapter opener in a local mathematics textbook to illustrate the use of the concepts of ratio in daily lives (Toh, 2012).

Figure 1. Chapter opener in a local textbook (Toh, 2012, p. 107)

Teachers could use this opportunity to guide the students to make sense of the above information in relation to the concept of Ratio. The teacher could provide the necessary scaffold:

1. *It is stated that the full seating capacity of the bus is 30 adult passengers. How many seated children can the bus take?*

Hypothetical questions in this context can also be generated to challenge the students to stretch their thinking, such as:

2. *Suppose 4 adult teachers are in the bus. How many children can this bus now take?*

There are many such real-life mathematical information in the real world, newspapers abound with such data. Using such real world mathematics situations to engage students in interpreting mathematical concepts in these situations would provide students with useful learning experience.

Making decisions in real-world context

Many real-world problems involve making decisions based on mathematical processes, or "heuristics". Providing students with ample opportunity to make decisions using mathematics in real-life situations can be seen as the first step to enable students to appreciate mathematics (Toh, 2010). In fact, it is not far-fetched to say that such learning experiences are the forerunner of introducing mathematical modelling to students.

Here one decision-making task will be demonstrated. Students first learn to perform arithmetical computation at the Lower Secondary level. Computation involving decimal numbers has been contextualized in monetary transaction since the early primary school math curriculum. Students have been given ample opportunity to perform such computation both manually and using calculators. Teachers could perhaps go one step further by tweaking some computation tasks into "authentic" tasks in which students are required to make decisions. An example of how a computation task is tweaked into a decision-making opportunity is illustrated in Figure 2.

Figure 2. Sample of a decision-making task involving arithmetical computation

In solving tasks involving decision-making, students must have the fundamental mathematical content and contextual knowledge. In addition to engaging the students to perform computation, teachers could invite the students to explain their reasoning behind their decisions (for example, choosing the lowest possible price). This could further provide students the opportunity to communicate and reason mathematically.

Decision-making tasks might not necessarily be a problem with multi-steps; it could be one involving a simple step followed by an interpretation as in Figure 2. It is not difficult for teachers to modify existing computational tasks into activities which require students to make decisions, or create such activities from real world scenarios, backed by mathematics concepts.

3.2 *Principle Two*

Students must be able to obtain satisfaction from their learning experience, in addition to building up their particular mathematical concepts or problem solving ability. Such satisfaction comes from providing students the opportunity to appreciate the real life application of mathematics, as illustrated in the preceding paragraphs. Another dimension of satisfaction comes from enabling students to experience the "power" of mathematics (Toh, 2011).

Pattern recognition is one core learning experience that students must acquire. This is echoed in many national curricula. For example, the National Council of Teachers of Mathematics (NCTM) is of the opinion that experiences of generality and recognition of general relationships in spatial patterns and number patterns are crucial in developing students' algebraic thinking. In the Singapore mathematics curriculum, students must be given sufficient opportunity to practice pattern recognition in preparation for their learning of algebra at the Lower Secondary level. It is not surprising that educators recognize pattern recognition as a critical success factor in mathematics (for example, Quinn, 2005).

Pattern recognition develops a child's ability to identify and describe attributes of objects and the similarities and differences between them. The use of pattern recognition to facilitate an individual to "discover new formula" will be demonstrated next. In fact, it is the most natural process of a working mathematician towards discovering new identities. To enable students to experience the "power" of pattern recognition in this aspect, students must be given such opportunities instead of merely completing the blanks of typical examination-type questions. An example of a task in which students discover that the sum of first n odd numbers is a perfect square is shown in Figure 3.

a) Fill in the blanks below:

$1 + 3 =$ _____

$1 + 3 + 5 =$ _____

$1 + 3 + 5 + 7 =$ _____

$1 + 3 + 5 + 7 + 9 =$ _____

b) What do you observe about the sums of the above series?

c) What is the sum of the first 1000 odd numbers?

d) Find the sum $1 + 3 + 5 + 7 + 9 + \ldots + 2009$.

e) Find the sum $1 + 3 + 5 + 7 + \ldots + (2n - 1)$ in terms of n.

Figure 3. Sample of a pattern recognition task

In the task in Figure 3, students could experience the satisfaction of having discovered something in mathematics – that the sum of the odd numbers is a perfect square! A teacher could skilfully lead the students to give a "plausible" explanation as to why this is so – is such a "nice" formula a mere coincidence or could it be more than this?

It is important that such experience should be included in the curriculum to enable a learner to feel the "reality" of mathematics and obtain the satisfaction that one has completed a worthy task. It could further develop students' curiosity in the subject to find out more. Curiosity is a motivationally original desire to know (Schmitt & Lahroodi, 2008) which sustains one's attention and interest to know more. It is an important link to an individual's lifelong learning (Fulcher, 2008). Curiosity in mathematics can lead students to explore new ideas in the subject (Gough, 2007). Curiosity is one attribute that is often observed in our students but which teachers usually ignore, overshadowed by the content-heavy curriculum to complete for the high-stake national examinations.

3.3 *Principle Three*

With careful planning, it is not likely that an experienced teacher would end up designing learning experiences that try to elicit responses from students and which are beyond their means. However, anecdotes can be found in actual classrooms, especially with the use of technology.

As an illustration, the use of Information and Communication Technologies (ICT) provides students with opportunities to explore mathematics through the intuitive-experimental approach prior to deductive approach of learning mathematics. The Singapore mathematics textbooks are replete with activities that allow students to use ICT to explore mathematics. In such situations, it is always important for the teachers to be mindful of the prior knowledge students have acquired about technologies in other school subjects. For example, if a spreadsheet is expected to be used to explore graphs and number patterns, or a dynamic geometry software to explore properties of geometrical figures, teachers must ensure that their students have already acquired knowledge on the use of the related software before engaging them in these activities.

In addition, teachers must always be mindful that the main instructional objective of such learning experiences is for students to learn mathematics, instead of the use of technology per se. As an example, when a teacher intends to use a dynamic geometry software to explore geometrical properties, it is likely that a well-designed template with clear instructions be made available for students to click and drag the icon on the computer screen to explore properties, rather than teachers expecting their students to perform construction of geometrical figures using the software. If much time needs to be spent to teach students to perform construction instead of pedagogy, this goes against the curriculum objective.

3.4 *Principle Four*

According to Tyler (1946),

> there are many particular experiences that can be used to attain the same educational objectives. As long as the educational experiences meet the various criteria for effective learning, they are useful in attaining the desired objectives. (p. 67)

It is likely that a particular learning experience or activity is more suitable for a particular group of students than the others. Teachers would then need to choose the most appropriate learning activity for their students. Generally, there are learning experiences which are more illuminating for a particular situation than the others, while there are learning experiences that are transferrable to other learning situations in addition to being effective for learning a particular mathematical concept.

In this section three methods that are commonly used in teaching lower secondary mathematics students about the sign of the product of two negative numbers will be demonstrated.

Method 1: Using analogy
The product of two negative numbers being positive is by no means intuitive to most students. To begin with, it is difficult to illustrate negative numbers, which are abstract to many young children. Number lines have been coined to represent negative numbers, but it is not clear how the product of two negative numbers can be represented. Some teachers skilfully invented creative mnemonics to help their students to memorize the fact. The method is like this: in the product of two numbers, let the first number be analogized to either good people (+) or bad people (-); the second number be entering your country (+) or leaving your country (-). Common sense reasoning could lead to the following diagram.

	Good people (+)	Bad people (-)
Enter your country (+)	Good (+)	Bad (-)
Leave your country (-)	Bad (-)	Good (+)

So bad people (-) leaving your country is good (+), hence the product of two negative numbers is positive.

In the above illustration, there is no logical reasoning as to why the above statements represent the product of two numbers, e.g. why "bad people leaving your country" should be represented by "negative x negative" instead of "negative + negative".   Nevertheless, teachers have shown that this is a rather useful mnemonics for memorizing the sign of the product of two numbers. This is an example of a learning experience that is an alternative to rule learning. Feedback by Singapore teachers generally reveals that most students are satisfied by the above analogy. However, one should be mindful that this approach does not satisfy the curiosity of the more inquisitive students.

Method 2: Using number discs

The Singapore Ministry of Education introduces the use of algebra discs as a model to help students construct meaning and make sense of algebra symbols and processes. In the latest mathematics curriculum for lower secondary students, the algebra discs (or number discs, if algebraic letters are not used) have been incorporated as an opportunity to provide useful learning experiences to help students conceptualize the meaning of arithmetical operations involving negative numbers. It has been shown in local textbooks that number discs provide a clear visual representation of addition and subtraction of negative numbers.

This approach used in the Singapore curriculum to explain the product of two negative numbers, (-2) x (-3) will be demonstrated next.

Taking *a* x *b* to mean *a* groups of *b* objects, we interpret 2 x (-3) as two groups of -3. Hence it is clear that 2 x (-3) = -6, see Figure 4.

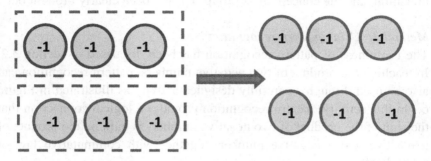

Figure 4. Use of number discs to demonstrate that 2 x (-3) = -6

Using the above definition of *a* x *b*, it would then mean that -2 x (-3) would be interpreted as -2 groups of -3. The negative sign of -2 indicates a flip of all the discs, resulting in Figure 5.

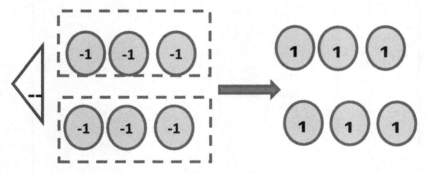

Figure 5. Use of number discs to demonstrate that (-2) x (-3) = 6

In the above illustration, -2 groups is used to denote 2 groups followed by a flip. Thus, negative sign in the above use of number discs is translated into a process of flipping the discs. The advantage of this method is that it builds on the elementary concepts of multiplication by introducing an additional flipping (which denotes negative numbers). It should be noted that the *concept* of multiplication of negative numbers has yet to be revealed by this method too. In this method, the multiplicative inverse is interpreted by the process of flipping the discs. However, to "force" the meaning of -2 groups of objects is rather unnatural, and the concept of -2 groups has not been clearly brought out.

Method 3: Using pattern recognition

The usefulness of pattern recognition has been discussed in section 3.2. In teaching the product of two negative numbers, pattern recognition can also be used through a carefully designed activity as illustrated in Figure 6. In this activity, pattern recognition provides a logical connection that the sign of the product of two negative numbers is built on the product of a positive and a negative number, together with commutative law of multiplication.

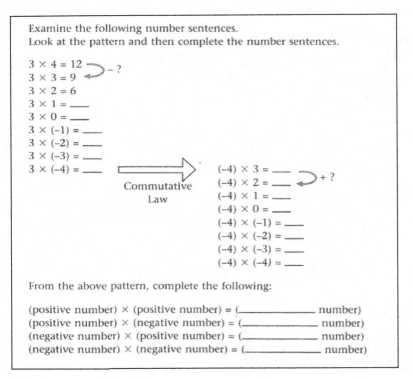

Figure 6. A pattern recognition task (Yeap, 2009)

What is noteworthy is that pattern recognition illustrated in Figure 6 is easily accepted by students (with some mathematical logic connection across concepts), and is *transferable* to other situations (Toh, 2011). We will illustrate with another situation that an activity similar to Figure 6 can be used to introduce zero and negative indices, which is abstract to lower secondary students.

How can we achieve an intuitive meaning of zero and negative indices, say of 2^0 and 2^{-3}? Pattern recognition can be used to give a fairly satisfactory answer to this question to avoid theoretical approaches. Consider the activity designed in Figure 7. By completing the first four lines and observing the pattern formed by the equations above it, students can accept that $2^0 = 1$ and $2^{-1} = \frac{1}{2}$ and so on.

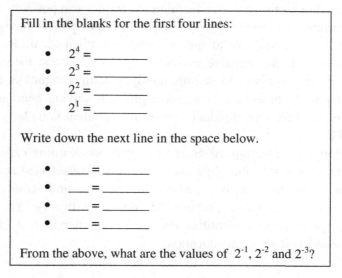

Figure 7. A proposed activity using pattern recognition

Readers should also be mindful that although pattern recognition as described above is useful for teaching such abstract ideas, they still have not met the mathematical rigors. In the words of Wu (2011), what is revealed above is still "half-truth" and has not met the precision required for mathematics. Interested readers are encouraged to read Wu (2011) for a detailed discussion.

There are many learning experiences that could be used to achieve the same educational objectives. It is important that teachers consider the advantages and disadvantages of each of the learning experiences and activities, and how these opportunities could clearly illuminate the mathematical concepts or could be generalized to other learning opportunities. Underlying the design of appropriate activities, teachers must have good mathematical content knowledge, pedagogical content knowledge and knowledge of their students.

3.5 *Principle Five*

As illustrated in section 3.4, one learning experience can possibly lead to several learning objectives. It was seen precisely that pattern recognition can be used to lead students to appreciate more mathematical ideas such as the product of two negative numbers is positive, or the meaning of zero and negative indices. More importantly, pattern recognition, if used consistently with students, is a valuable problem solving heuristic that can be a method for an individual to learn new content knowledge (Toh et al., 2011).

Educators have also tapped on the fact that one learning experience can lead to several learning objectives. Teachers have selected activities that demonstrate the use of numbers in real life and stimulatingly sensitize students to the importance of Singapore history. The same learning experience that identifies the use of number in real life can sensitise students to national education.

4 Organization of Learning Experience

One should note that no single learning experience has a profound influence on a student nor can it cause a change in his or her behaviour within a short time. Any desirable changes in behaviour can only take place after a period of time with an accumulation of a systematically organized set of learning experiences. For effective teaching and learning, Tyler (1946) offered three criteria for organization of these selected learning experiences:

- Criterion One – Continuity: there must be recurring opportunity for required skills and knowledge to be practiced and developed.
- Criterion Two – Sequence: subsequent experiences must be built on earlier ones, more in depth but not just mere repetition.
- Criterion Three – Integration: the learning experience must help students to develop a unified view in relation to the elements dealt.

The next section demonstrates how learning experiences related to a specific topic of solving linear algebraic equations can be organized for effective instruction, taking the above material into consideration.

4.1 *Planning of learning experience: An example of solving linear algebraic equations*

Algebra is generally difficult for students, partly because it is dominated by symbols. It is a common observation among mathematics teachers that students possess procedural knowledge associated with algebra that is not backed by conceptual understanding (Yeap, 2009). Unfortunately, such procedures are further reinforced by the ways teachers teach in classrooms and the words they used, such as "cancel" and "bring over" in solving algebraic equations (Martinez, 1988).

In this section, it will be demonstrated how the learning experiences associated with algebraic equations can be organized into an organic whole to enhance meaningful learning of the entire topic. These are different ways to conceive algebraic equation:

- The missing numbers in a riddle.
- The missing numbers in an equation.
- Balancing two sides of a beam balance.

Students should be provided with learning experiences associated with all the above conceptions, moving from one stage to the next – from the more intuitive to the more sophisticated one.

Stage 1: Riddles as brain teasers

It is very useful to begin with riddles or guessing games, which many students already had been exposed to since their primary school education. Teachers could design simple puzzles and invite them to participate and even create their own puzzles to challenge their peers, an example of which is shown in Figure 8.

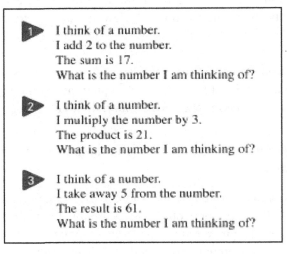

Figure 8. Sample of a number-guessing game

Riddles and guessing games are useful in capturing the attention of students to the related mathematical tasks. In the above example, the riddle serves as a prelude to solving algebraic equations. Students can solve such problems by basic guess-and-check – a problem solving heuristic that they have acquired in primary school education. This provides the continued opportunity to practice the heuristic. Students may be enthused to "discover" quicker methods to solve the riddle, especially when the answers become complicated – teachers can skilfully link this to the procedure of solving an algebraic equation. At this stage, students are not exposed to the physical form of an algebraic equation yet.

Stage 2: Finding the missing numbers in an equation involving numbers
At the next stage, teachers can lead students to experience the process of
"guess-and-check" in an equation involving numbers and the four
operations. A sample of such activities is shown in Figure 9, which
consists of three arithmetic equations with some numbers smeared and
hence not readable. While the technique of guessing the answers is fairly
similar to Stage 1, the physical appearance of the equations in this figure
resembles that of an algebraic equation. From stage 1 to stage 2, students
must be given ample opportunity to explore and appreciate the ideas of
solving an equation. Teachers should not plunge immediately into the
abstract procedure of solving equations.

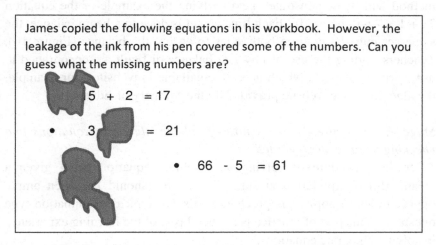

Figure 9. Sample of a guessing activity in teaching algebraic equation

Students could then be challenged if they could think of an efficient
method to find the unknowns. Teachers are then ready to introduce the
use of algebraic symbols to represent unknown numbers in each of the
equation as an economical form of representing such equations. For
example, the three equations in Figure 9 can then be written in the form

- $x + 2 = 17$
- $3 \times y = 21$
- $z - 5 = 61$.

Stage 3: Solving algebraic equations as an act of balancing

After students start to appreciate the use of letters in algebraic equation, the natural progression in teaching algebraic equations will then be to introduce the procedure of solving the equation. Literature suggests that students generally apply algebraic procedures that are not backed by conceptual understanding. Therefore, it will be useful to link the method of solving an algebraic equation to the act of balancing two sides of a beam balance. This provides a link between an abstract procedure and a concrete process, thereby facilitating students to memorize the procedure.

Solving the linear equation of the form $ax + b = c$ using the balance method will be demonstrated next (solving the example of the equation $2x + 1 = 7$ will be used). Each time an act is done to a beam balance, the same act must be done to both sides to ensure the balance (Figure 10). Teachers finding the use of physical balance difficult are encouraged to use "virtual balance", which is easily available in websites, for example, the algetool in the website provided by the Ministry of Education.

Stage 4: Applying the procedure of solving algebraic equations and checking the answers obtained

After the procedure of solving an algebraic equation being given a meaningful interpretation at stage 3, students should be given ample opportunities to apply this procedure to solve typical examination-type questions. This part of practice is a crucial part of the learning experience in solving algebraic equations.

To further reinforce the concept of an algebraic equation, students must be engaged in checking the correctness of their answers. This provides students the opportunity to check their answers, which is a crucial aspect of metacognition. It also reinforces the meaning of the solutions of an algebraic equation.

Balance Diagram	Equation
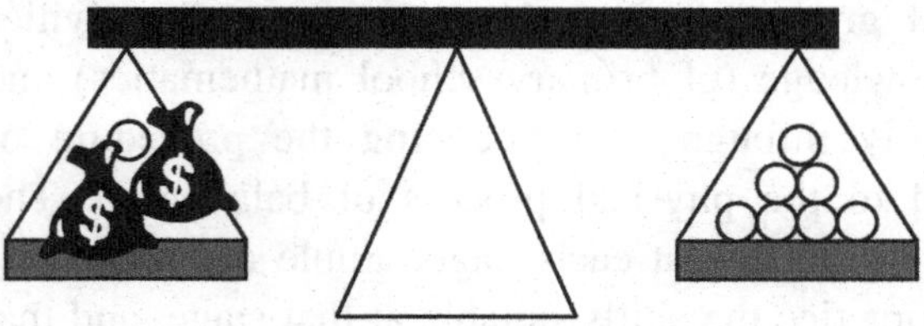	$2x + 1 = 7$
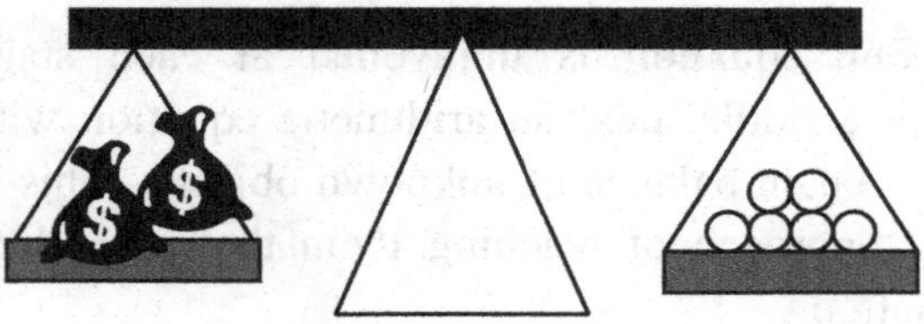 Remove one ball from each side of the balance	$2x + 1 - 1 = 7 - 1$ $\therefore 2x = 6$
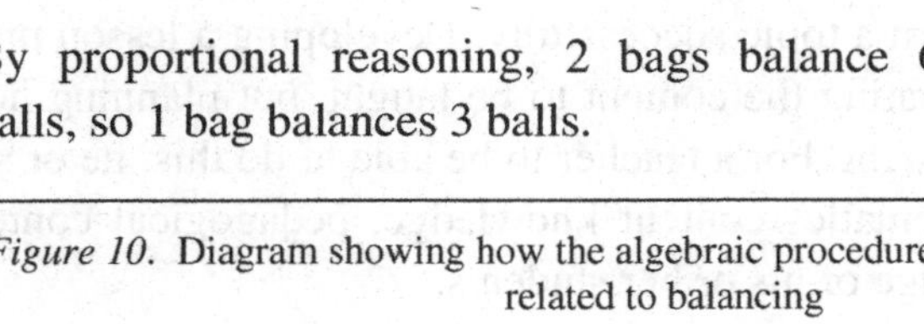 By proportional reasoning, 2 bags balance 6 balls, so 1 bag balances 3 balls.	$2x \div 2 = 6 \div 2$ $\therefore x = 3$

Figure 10. Diagram showing how the algebraic procedure of solving an equation is related to balancing

4.2 *Overview of the planning of learning experience*

In the above illustration of the planning of the learning experiences associated with solving a linear algebraic equation, the concept and processes are introduced gradually, through a sequence of activities linking to their prior knowledge (of primary school mathematics) and building on it sequentially through to introducing the procedure of solving, which is linked to the physical process of balancing. The experiences organized are such that at each stage, ample opportunity is provided for students to practice the skills suitable at that stage, and that students are ready to move to the next stage when they are familiar with the earlier ones.

The idea of an algebraic equation is unravelled at each stage sequentially, beginning with a riddle, next an arithmetic equation with missing numbers, and then a beam balance of unknown objects. This is contrasted to the "sudden" approach of teaching them the method of solving linear algebraic equation.

Through these stages, the students would have developed a clearer understanding and a unified view of algebraic equations even before they are able to solve such an equation formally.

5 Conclusion

In this chapter we have introduced the principles of selecting appropriate learning experiences and the organization of learning experiences to facilitate students to learn a topic successfully. Developing a lesson must involve not only deliberating the content to be taught, but planning how the content should be taught. For a teacher to be able to do this, he or she must have sound mathematics content knowledge, pedagogical content knowledge and knowledge of his or her students.

References

Barton, B. (1993). Ethnomathematics and its place in the classroom. In E. McKinley, P. Waiti, A. Begg, B. Bell, F. Biddulph, M. Carr, M. Carr, J. McChesney & J. Young-Loveridge (Eds.), *SAME papers 1993* (pp. 210 – 231). Hamilton: Centre for Science and Mathematics Education Research, University of Waikato.

Beswick, K. (2011). Putting context in context: An examination of the evidence for the benefits of 'contextualized' tasks. *International Journal of Science and Mathematics Education*, 9, 367 – 390.

Boaler, J. (1993). Encouraging transfer of 'school' mathematics to the 'real world' through the integration of processes and content; context and culture. *Educational Studies in Mathematics*, 25, 341 – 373.

Fulcher, K. H. (2008). Curiosity: a link to assessing lifelong learning. *Assessment Update*, 20(2), 5-7.

Gough, J. (2007). How small is a billionth? *Australian Mathematics Teacher*, 12(2), 10 - 13.

Hedberg, J., Wong, K. Y., Ho, K. F., Lioe, L. T., & Tiong, Y. S. J. (2005). Developing the repertoire of heuristics for mathematical problem solving: First Technical Report for Project CRP38/03 TSK. Singapore: Centre for Research in Pedagogy and Practice, National Institute of Education, Nanyang Technological University.

Lester, F. K. (1980). Research on mathematical problem. In R. J. Shumway (Ed.), *Research in mathematics education* (pp. 286-323). Reston, Virginia: NCTM.

Martinez, J. G. R. (1988). Helping students understand factors and terms. *Mathematics Teacher*, 81(9), 747 – 751.

Ministry of Education. (2012). *O-level mathematics teaching and learning syllabus*. Singapore: Author.

OECD. (2010). *PISA 2012 mathematics framework*. Paris: OECD Publications Retrieved March 6, 2014 from http://www.oecd.org/dataoecd/8/38/46961598.pdf.

Quinn, R. J. (2005). A constructivist lesson to introduce arithmetic sequence with patterns. *Australian Mathematics Teachers*, 61(4), 18 – 21.

Saugstad, T. (2013). The importance of being experienced: An Aristotelian perspective on experience and experience-based learning. *Studies in the Philosophy of Education*, 32, 7 – 23.

Schoenfeld, A. (1985). *Mathematical problem solving*. Orlando, FL: Academic Press.

Schmitt, F. F., & Lahroodi, R. (2008). The epistemic value of curiosity. *Educational Theory*, 58(2), 125-148.

Siu, M. K. (2001). History of mathematics in education – entrée? main course? or dessert? Unpublished Singapore Mathematical Society talk at the National University of Singapore 20 Oct 2001.

Taba, H. (1962) *Curriculum development: Theory and practice.* New York: Harcourt Brace and World.

Toh, T. L., Quek, K. S., Leong, Y. H., Dindyal, J., & Tay, E. G. (2011). *Making mathematics practical: An approach to problem solving.* Singapore: World Scientific.

Toh, T. L. (2010). Making decisions with mathematics: from mathematical problem solving to modelling. In Kaur, B., & Dindyal, J. (Ed.), *Mathematical applications and modelling: AME Yearbook 2010* (pp. 1-18). Singapore: World Scientific.

Toh, T. L. (2011). Exploring mathematics beyond school curriculum. In Bragg, L.A. (Ed.) *Maths is multi-dimensional* (pp. 77-86). Melbourne, Australia: Mathematical Association of Victoria.

Toh, T. L. (2012). *Math 360 Normal (Technical) for Sec 1.* Singapore: Marshall Cavendish Education.

Tyler, R. W. (1946). *Basic principles of curriculum and instruction.* Chicago: The University of Chicago Press.

Wu, H. (2011). Phoenix rising: Bringing the common core mathematics standards to life. *American Educator, 35*(3), 3 – 13.

Yeap, B. H. (2009). Teaching of algebra. In P.Y. Lee & N. H. Lee (Eds.), *Teaching secondary school mathematics: A resource book* (pp. 25 – 50). Singapore: McGraw-Hill.

Chapter 4

Providing Students' Authentic Learning Experience through 3D Printing Technology

Oh Nam KWON   Jee Hyun PARK   Jung Sook PARK

Authentic learning has been shown to help connect students' classroom learning to the outside world. Linking what students are learning in class to the real world enables them to better understand the problems they may face in real-world environments. Technologies can play an important role in supporting students' learning in an authentic environment. In this chapter, we explore the attributes of an authentic learning environment and 3D printing technology. Based on these attributes, we demonstrate how utilizing 3D printing technology to perform mathematics tasks gives students an authentic learning experience.

1 Introduction

Learning-by-doing is generally considered one of the most effective learning methods. But in mathematics classrooms in Korea, the emphasis in secondary schools has been on applying principles, concepts, and facts in an abstract and decontextualized form. The recent PISA 2012 result indicates that Korean students' attitudes towards mathematics are low when compared with the other 65 participating countries.

Authentic learning is a pedagogical approach that allows students to explore, discuss, and meaningfully construct concepts and relationships in contexts that involve real-world problems and in projects that are relevant to the learner (Donovan, Bransford, & Pellegrino, 1999). With the emergence of new technologies, we can offer students a more authentic learning experience, ranging from experimentation to real-world problem solving. Jonassen, Carr and Yueh (1998) have suggested that technology can become students' intellectual partner and help them analyse, synthesize and organize their knowledge and understanding. This is in line with how authentic learning can be incorporated into mathematical tasks using 3D printing as a technological support, since authentic learning requires learners to apply theoretical knowledge in a real context, and incorporating authentic learning strategies will have an impact on their cognitive, social and affective behaviours.

Although 3D printing technology has promising potential in mathematics education, there are few cases of it being applied to school education. In this chapter, the features of authentic learning and the potential of 3D printing technology are considered from the perspective of the learning experience in school mathematics education. In addition, we describe how the performance of tasks using 3D printing technology can contribute to students' authentic learning experience.

2 Authentic Learning Experience

Bridging the gap between the learning experience and the complexity of the real world is a constant challenge for educators. Though Dewey (1897) did not explicitly mention the word "authentic" in his book and papers, he seemed to put a strong emphasis on experience-based learning, and conveyed an equivalent set of beliefs shared by today's advocates of authentic learning: a constructivist point of view of creating personal meaning out of real events and interaction. This was a century ago.

Piaget and Inhelder (1974) believed that the learner must be active to be engaged in real learning. Learning becomes active when students are able to connect new knowledge with their prior comprehension. Constructivists took this notion a bit further, stating that a meaningful context that brought the real world into the learning environment was key to promoting learning (Brown, Collins, & Duguid, 1989). Learning is a process of interacting with the outside world, and continually reanalysing and reinterpreting new information and its relation to the real world (Brown et al., 1989; Lave & Wenger, 1991). Traditional learning situations in which students are passive recipients of knowledge are inconsistent with the learning situations of real life (Lave, 1988).

In the 21st century, more relevant, authentic and applied teaching and learning strategies need to be incorporated into learning environments to innovate the student learning process. To make student learning relevant to real life experiences, learning environments must be authentic.

2.1 *What is authentic learning?*

Authentic learning is based on situated learning theory, which was founded on a constructivist philosophy of learning. It is a process involving dynamic interactions between the learner, the task and the environment. Authentic tasks that encourage and support student engagement and immersion in a cognitive real environment can facilitate self-directed and independent learning (Herrington, 2006).

Authentic learning is defined as learning that focuses on real-world, complex problems and their solutions, using role-playing exercises, problem-based activities, case studies, and participation in virtual communities of practice (Lombardi, 2007). Authentic learning is important because it allows students to experience real-world problems while in a learning environment (Herrington et al., 2004; Herrington & Herrington, 2006), and to combine concepts and theories from formal education and apply them to real world practice (Bennett, Agostinho, & Lockyer, 2005; Borthwick et al., 2007).

The authentic learning environment is inherently multidisciplinary (Lombardi, 2007). It intentionally brings into play multiple disciplines, multiple perspectives, ways of working, habits of mind, and community. It can be useful for a multidisciplinary approach in science, technology, engineering, art and math (STEAM) education.

Above all, in an authentic learning environment students are motivated by performing actual work tasks, and create products that reflect who they are and what they believe in (Wagner, 2008). Researchers had developed a theoretical or practical perspective for authentic learning, with implications for higher education (Borthwick et al., 2007; Elliot, 2007; Herrington & Oliver 2000; Herrington, Oliver, & Reeves, 2003; Harrington et al., 2009; Meyers & Nulty, 2009).

According to Borthwick et al. (2007), there are three common types of authentic learning models. They all share a common origin in the notion of situated learning. The models are as follows:

- Apprenticeship model: students are mentored by a professional who provides an authentic work experience in the real world.
- Simulated reality model: learning activities are engaged in that seek to simulate the "real world".
- Enminding model: connection between students' experiences and disciplinary "mind" through authentic activities.

To develop students' mastery of spatial sense and mathematical skills, the enminding model offers the best alternative for this study. In our project, students were required to perform tasks that linked their experiences and their mathematical knowledge through authentic activities. In this project, we provided students with the views of experts, and simulated real-world situations using 3D printing technology. But the ultimate aim was to develop students with analytical, critical, and reflective mathematical skills. The enminding model offers a more appropriate approach because it requires engagement with the discipline as an authentic activity.

Herrington and Herrington (2006) described the general principles and characteristics for designing an authentic learning environment for higher education. They identified 9 characteristics of authentic learning based on constructivist philosophy and approaches, situated learning theory in particular. Their work provides educators with the following useful checklist, which can be adapted to any subject.

- Authentic contexts
- Authentic activities
- Access to expert performances and the modelling of processes
- Multiple roles and perspectives
- Collaborative construction of knowledge
- Reflection
- Articulation
- Coaching and scaffolding

- Authentic assessment.

This framework has been used to design and/or evaluate a number of technology-based learning environments on the theoretical foundation of authentic learning. The project to be shown in this chapter was also analysed based on this framework.

2.2 Why is authentic learning important in the mathematics classroom?

Blum (1993) argued for the importance of situated authentic learning in mathematics teaching and learning, based mainly on the general goals and aims of mathematics instruction which were as follows.

Pragmatic aspect
Teaching mathematics is intended to help students to understand and to deal with real-world problems and situations. As such, authentic learning is indispensable.

Formative aspect
Students can acquire a general literacy that goes beyond mathematical literacy, and develop an attitude of openness to new situations.

Cultural aspect
Students should learn mathematics topics as a source for reflection, or to build a comprehensive and balanced picture of mathematics as a science and as a part of human history and culture. Authentic learning is an essential feature of human intellectualism as well as of history and actual practice, and can thus contribute towards promoting those aspects.

Psychological aspects
Mathematical contents can be motivated or consolidated by a suitable authentic learning experience, and these may contribute to a deeper understanding and longer retention of mathematics topics, or they may improve students' attitudes toward mathematics.

Students very often experience mathematics as the mechanical manipulation of meaningless symbols. An authentic learning experience can contribute towards giving more meaning to the teaching and learning of mathematics. If the formation of appropriate basic ideas and the development of meaning are essential aims of mathematics instruction, then authentic learning must be an integral part.

Recently, the Korea Foundation for the Advancement of Science and Creativity (KOFAC) developed STEAM-related intensive courses for high school students in science and technology with a view to their future academic careers. They also emphasized application-oriented activities in the mathematics classroom. Authentic learning has the potential to improve student engagement in these courses, and educational outcomes.

3 3D Printing Technology

In 21st century workplaces, students can build their digital literacy and other sophisticated skills for the global, knowledge-based, innovation-centred economy (Wagner, 2008). The New Media Consortium (NMC) (2013) is an international not-for-profit consortium of more than 250 colleges, universities, museums, corporations, and other learning-focused organizations dedicated to the exploration and use of new media and new technologies. Horizon Report (NMC, 2013), the annual report published by NMC, is an unbiased source of information that helps education leaders, trustees, policy makers, and others easily understand the impact of key emerging technologies on education, and when they are likely to enter mainstream use. According to the 2013 higher education edition of Horizon Report, 3D printing will be popular in four or five years time.

3.1 *What is three-dimensional (3D) printing?*

The industry of rapid prototyping and 3D printing in particular emerged about 30 years ago and is considered by some to be part of an industrial revolution in which manufacturing becomes digital, personal, and affordable (Knill & Slakovsky, 2013). First commercialized in 1994 with printed wax material, the technology has moved to incorporate other materials like acrylate photopolymers or metals, and is now entering the range of consumer technology. The development of 3D printing is the latest piece in a chain of visualization techniques. These changes also affect mathematics education.

3D printing involves technologies that construct physical objects from three-dimensional (3D) digital content such as computer-aided design files (CAD), computer aided tomography (CAT), and X-ray crystallography. This is known in the industry as rapid prototyping. A 3D printer builds a tangible model or prototype from the electronic file, one layer at a time, using an inkjet-like process to spray a bonding agent onto a very thin layer of fixable powder, or an extrusion-like process using plastics and other flexible materials.

The deposits created by the machine can be applied very accurately to build an object from the bottom up, layer by layer, with resolutions that are more than sufficient to express a large amount of detail. The process even accommodates moving parts within the object. Using different powders and bonding agents, colour can be applied, and prototype parts can be rendered in plastic, resin, or metal. This technology is commonly used in manufacturing to build prototypes of almost any object that can be conveyed in three dimensions (NMC, 2013).

3D printing is quickly becoming a very affordable option for producing physical objects. 3D printing encompasses a number of closely related technologies, all of which produce a 3-dimensional physical object from a computer model by building it up in successive layers. It allows a designer to convert a computer model of a prototype into a physical object quickly in comparison to most previously available technologies (Segerman, 2012).

Today, 3D graphics programs are also widely available. Anyone with a computer can afford the technology to design in three dimensions. Google SketchUp or Blender can be downloaded for free. As well, many jobs now require more 3D skills, including jobs in medicine, mining, design, and countless other fields.

3.2 *What are the benefits and difficulties of using 3D printer in a mathematics class?*

There are various reasons why schools should seriously consider using 3D printing in the classroom (Cox, 2011; Kross, 2011; Wolf, 2013). These include the following benefits:

A 3D printer encourages the imaginations of students
One of the challenges mathematics teachers face is the difficulty in inspiring students to understand and study various 3D figures when they never have the chance to actually see, hold, touch or feel their designs as a final physical product. For many students, the models they design on the computer screen are perceived as unobtainable fantasy objects. A 3D printer can take the 3D models the students design with computer aided drafting software and create plastic versions. This is an important advantage that some schools are already offering.

3D printers result in tangible objects for many students
3D printers are good for the school and are a technology that can also aid and integrate various other disciplines. For example, students of architecture can quickly make physical examples of their designs; students of science can make three-dimensional molecular models, and fine art students can 3D print real-life examples of their designs. Students in STEAM education can combine mathematical concepts with scientific or artistic experiences and objects using 3D printing (Wolf, 2013). With the printer to carry out the production of objects, more time can be spent considering the science and mathematics involved in design.

A 3D physical model can aid students' spatial reasoning
3D printers can aid in the development of advanced spatial reasoning capabilities. Something as simple as the act of rotating and observing a prototype can have a profound effect on a student. Spatial intelligence is the capacity to mentally generate, rotate, and transform visual images for spatial judgment, and the ability to visualize with the mind's eye (Gardner, 2006). It is the ability to draw accurate conclusions from observing a three-dimensional environment. It involves interpreting and

making judgments about the shape, size, movement, and relationships between surrounding objects, as well as the ability to envision and manipulate 3D models of things that are not immediately visible.

There are a number of advantages that make 3D printing attractive for making mathematical models. One is that there is a huge amount of freedom in the geometry that can be produced. With subtractive manufacturing techniques, in which material must be removed from an initial solid (e.g. lathing, drilling or carving), objects with an intricate internal structure can be very difficult to produce. With 3D printing's additive procedure, these problems are greatly mitigated.

3D printers introduce students to modern manufacturing processes
3D printing is a cutting-edge technology, and is one of the modern production techniques that it is essential for schools to educate students in. It is a great way to teach students about how manufacturing works. We want to see more students have experience that leads to a career in STEAM, and help educators see the importance of 3D printers as another tool in STEAM education. 3D printers have value for classrooms and businesses alike because of their durability and performance.

Despite these advantages, most of the programs used to 3D print are still focused mainly on the design and engineering market, and are still high in terms of cost.

3.3 *How does 3D printing technology support an authentic learning experience?*

It's not easy to set up an authentic situation in a learning environment. Technology can provide access to software visualizations, images, and audio, bringing abstractions to life.

One of the most significant aspects of 3D printing for education is that it enables a more authentic exploration of objects. NMC (2013) reported that while 3D printing is four to five years away from widespread adoption in higher education, it is easy to pinpoint the practical applications that will take hold. The exploration of the 3D printing process from design to production, as well as demonstrations and participatory access, can open up new possibilities for learning activities.

Harvard University's Semitic Museum is already using 3D printing technology to restore a damaged ancient artefact from their collection (Time, 2013). By 3D scanning existing fragments of an Egyptian lion's legs, researchers are able to create computer models that will be used to print a miniature of the complete sculpture, though it initially was missing its body and head. Hart (2005) has written previously in the Mathematical Intelligencer on 3D printed mathematical visualizations, covering models of the Sierpinski Tetrahedron and the Menger Sponge, as well as various polyhedral designs and projections of 4-dimensional polytopes.

3D printing integrates many different disciplines including computation, geometry, art, physics, and others. The opportunity for students to design objects using a variety of skills is limitless. Teachers should be looking for ways to integrate curriculum and lessons using 3D printing as a way to motivate students, and to create opportunities for authentic educational activities.

4 The Project of Restoring Cultural Artefacts

The project was carried out with high school gifted students in a mixed first and second grade integrated mathematics and science class over a series of two 100-minute lessons. This project involved an authentic situation, linking abstract mathematical concepts with an archaeological situation. The key mathematical concepts related to a quadratic curve in two dimensions, and a curved surface in three dimensions. The required technical skills were related to graphics, design and 3D printing using the mathematics technological tool. The aim of the project was to help students develop the ability to apply these mathematics concepts to solving problems with technological skills, and to experience authentic learning in a real situation. It consisted of two phases: exploring Jeulmun Pottery from both a socio-cultural and a mathematical perspective, and reproducing it with 3D printing technology.

4.1 *Phase 1 exploring Jeulmun pottery*

The Jeulmun Pottery Period is an archaeological era in Korean prehistory that dates to approximately 8000-1500 BC. (Crawford & Lee, 2003). Jeulmun means 'Comb-patterned'. This comb-patterned pottery has a different shape from pottery found today. The Jeulmun pottery is V-shaped, with a pointed or rounded bottom.

In Task 1, students were asked to explain why the people of the time made a simple V-shape for the Jeulmun pottery (Figure 1). If they understood the lifestyle at that time, then they would know the reason for such a shape.

Figure 1. Jeulmun pottery

In Task 2, students were asked to two-dimensionally explore the shape of the pottery in order to reproduce the pottery as it would have been in Jeulmun Period. Students were provided with a reduced photo of pottery to complete this task. Students needed to explore the shape of pottery using mathematical concepts and formulas inside a realistic and cultural context (Figure 2).

Task 2. One day, your colleague Suyun finds the broken pieces of Chulmun pottery in Amsa-dong, Seoul. You are trying to help restore and reproduce the vessel using 3D printing technology.

1) Find a functional formula fitting the outline curve of the complete Chulmun pottery using dynamic geometry software (eg. Geogebra and GSP, etc.)

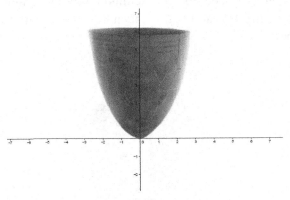

2) Find the formula representing the surface of revolution generated by revolving the above curve about the y-axis.

Figure 2. Exploring the Chulmun pottery mathematically in Task 2

4.2 *Phase 2 Reproducing Jeulmun pottery with 3D printing technology*

The rotation of a parabola about its axis forms a paraboloid. Unlike Task 2, which was focused on 2-dimensional exploration, in Task 3, students were asked to design and to print the model of Jeulmun pottery using a 3D printer based on the result of their exploration. To achieve this task, students could use mathematical or design software to construct the model. When they wanted to know how to use this technology, an instructor provided the relevant lessons. Those lessons taught them the simple skills and offered the resources required to deal with the software, such as Sketchup, Mathematica or Matlab. The tasks were performed for about 2-3 weeks.

We will describe the activities of the project that attempted to link theory with practice to provide an authentic learning experience based on the attributes of authentic learning identified by Herrington in the following.

4.3 *Designing an authentic learning environment*

Provide an authentic context and tasks that reflect the manner in which the knowledge will be used in real life

Herrington & Herington (2006) stated that the context needs to be all-embracing, to provide the purpose and motivation for learning, and to provide a sustained and complex learning environment that can be explored at length.

The context of the projects was restoring or reproducing cultural assets. Today, it is possible to effectively extract inner patterns and silhouettes from data by using 3D printing technology. 3D printing requires the management of 3D vector data, so mathematical knowledge is necessary to use 3D printing technology. The project began with restoring a cultural asset. Students explored their methods of extracting data from the damaged cultural assets, and built a model for reproducing the assets.

It is not enough to provide suitable examples from real-world situations to illustrate the concept or issue being taught. It encompasses a physical environment that reflects the way the knowledge will be used, and a large number of resources to enable sustained examination from different perspectives. The context provides a realistic and authentic rationale for the study of complex problems, a 3D-model reproduction that is relevant to the real world.

The tasks were related to understanding quadratic curve and the visualization of a three-dimensional figure. However, this knowledge was offered in the real context of restoring artefacts. The tasks were comprised of ill-defined activities that have real-world relevance.

Access to expert's performances and processes of modelling

Exposing students to expert performance gives them a model of how a real practitioner behaves in a real situation. Access to such modelling of processes has its origins in the apprenticeship system of learning, where students and craftspeople learned new skills under the guidance of an expert (Collins, Brown, & Newman, 1989).

Students were given the examples of an expert's task performance, or of an expert's comment, to enable them to model real-world practice. The samples of pottery model printing could be available to students (Figure 3).

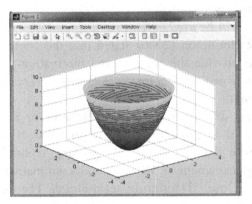

Figure 3. Expert's designed sample using Matlab

Figure 3 shows the example designed with Matlab. When students wanted to use Mathematica, the instructor provided the basic code required to generate the graphic output. Students could design a 3-D model based on the example.

Multiple roles and perspectives
Task 2 was relatively undefined and open to multiple interpretations, requiring students to identify the solution for themselves. Students attempted to solve the task using the multiple approaches such as the analytic, geometric and statistical ones.

Some students tried to fit the quadratic functional formula induced from their conjectures (Figure 4). Other students tried to draw the parabolic curves using the definition as the path (locus) of a moving point so that its distance from a fixed line (the directrix) was equal to its distance from a fixed point (the focus) (Figure 5).

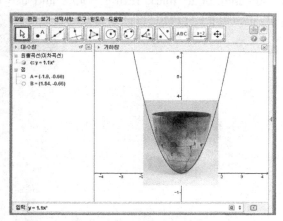

Figure 4. Students' analytic approach to Task 2

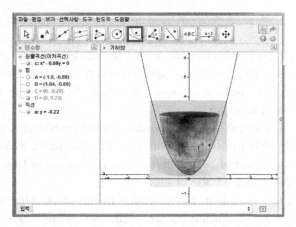

Figure 5. Students' geometric approach to Task 2

The vertex of the parabola is the point on the curve that is closest to the directrix; it is equidistant from the directrix and the focus. Thus, students first drew any line parallel to the x-axis as directrix and then put the vertex and the focus to determine a line perpendicular to the directrix. That line was the axis of the parabola. The parabola was symmetrical about its axis, moving farther from the axis as the curve recedes in the direction away from its vertex.

There was an attempt to mark fitting points and use the quadratic regression analysis using a spread sheet (Figure 6).

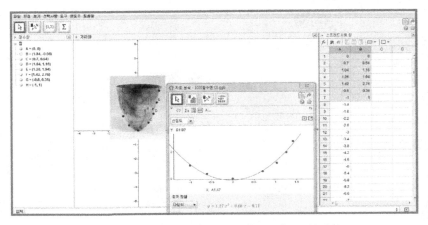

Figure 6. Students' statistic approach on Task 2

Students used the regression curve and found the functional formula to fit the picture. Their various approaches to solving the task were discussed through the whole class discussion from the perspectives of effectiveness and accuracy.

In general, content tends to be discipline-specific and presented in sections, with little to offer students seeking alternative viewpoints. But providing a multitude of perspectives is more conducive to sustained and deep exploration of any issue or problem.

The approaches taken by students to the task showed that the authentic learning environment enabled and encouraged students to explore different perspectives.

Collaborative construction of knowledge
Collaboration has been defined as the mutual engagement of participants in a coordinated effort to solve a problem together (Roschelle & Behrend, 1995). Forman and Cazden (1985) have suggested that true collaboration is not simply working together but also involves solving a

problem or creating a product that could not have been completed independently.

The activities were designed to provide students with opportunities to collaborate and reflect socially on the project. Activities and problems can be addressed to a group. Collaboration can be encouraged through complex tasks and technology (Figure 7). Group discussions and whole discussions can be used to encourage the sharing of ideas and joint problem solving within and among groups.

Figure 7. Collaboration using a computer for Task 2

Reflection & Articulation

In these activities, students were required to articulate and justify their work to their peers. Students were required to present and defend their arguments, through group discussion or whole class discussion.

When students approached a new subject in mathematics for the first time, they immediately found it difficult to perceive the relevance of the new concept to their life experience. The tasks provided the chance for the learner to look for a connection between mathematics and other subjects in new contexts. Those contexts needed to be explored. The concepts being learned are always part of a much larger learning. Students had to apply the concepts being learned, to link them and to articulate them.

Authentic and meaningful activities could promote reflection and allow the students to compare their opinions with experts. Students reflected on their solutions through discussing them with their peers.

Collaborative groupings enabled students to reflect socially and to engage in meaningful discussions on the issues presented. They discussed the potential and universalization of 3D printing technology in the future as well as their 3D restoring method.

Writing reflective journals provided a tangible outcome of students' reflections. Speaking after the experience, students mentioned that they had an opportunity to rethink a three-dimensional figure in a real-world context, and had helped to design a model of pottery. Figure 8 shows an excerpt from a student's journal.

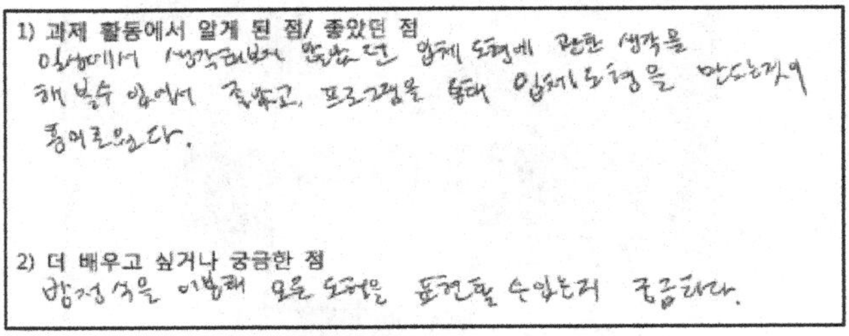
1) 과제 활동에서 알게 된 점/ 좋았던 점
일상에서 생각해보지 않았던 입체 도형에 관한 생각을 해 볼수 있어서 좋았고, 프로그램을 통해 입체도형을 만드는것이 흥미로웠다.

2) 더 배우고 싶거나 궁금한 점
방정식을 이용해 모든 도형을 표현할 수 있는지 궁금하다.

Figure 8. Example of student's reflective journal

In his reflective journal, the student wrote that the task gave him a chance to think about three-dimensional concepts in a real-world context, and that it was very interesting to construct three-dimensional objects using computer software.

Coaching and scaffolding

In many cases, the teacher's role was a didactic one, 'telling' students what they need to know rather than serving in a coaching role. A more authentic environment such as this project provided for coaching at critical times and scaffolding of support. The teacher and/or student peer mentors provided the skills, strategies and links when the students were unable to perform the task.

In Task 3, students needed to understand the shape of a three-dimensional space. Students could use the Cartesian coordinates directly or the parametric equations. At that time, the teacher could even surrender some of his or her own power as an expert to join students as co-learners. Since the students were not in the same grades, collaborative learning with more able partners could assist with scaffolding and coaching, as well as provide the means for the teacher to support their learning.

In this project, students could be supported in their innovation through visualizations, simulations, and interactive technologies, including the program for 3D printing. They also were provided various resources to explore.

Authentic assessment

In this project, the tasks functioned as a kind of assessment itself. They were integrated with the activity-based real situation and provided the opportunity for students to be effective performers with acquired knowledge, and to craft products or performances in collaboration with others. Many of the researchers and teachers exploring the model of situated learning had accepted that the technology could provide an alternative to the real-life setting (Herrington, Oliver, & Reeves, 2003). These tasks would also be offered in a well-designed on-line environment.

In any case, technology helped students to conjecture and check their assumption in this project. This is thanks to the fact that technology enables immediate feedback and interactivity. Software to mathematically explore artefacts and 3D printing technology supported their problem solving in an authentic context. Students experienced real work related with archaeology and represented their mental figure related mathematics.

In this project, 3D printing technology opened up possibilities for innovative and effective teaching-learning opportunities.

5 Conclusion

We illustrated how the project met the requirements of an authentic learning experience using the characteristics identified by Herrington (2006). We found how 3D printing technology offered students' authentic learning experience in the context of application-oriented and multidisciplinary curriculum.

5.1 *What did students experience in terms of learning?*

Generally, students showed a high interest in and enjoyed learning authentic tasks. The students also demonstrated an awareness of their classroom learning environment by being able to provide practical suggestions on how the activities could be made more exciting and relevant, such as through the use of technology, group work, or project work. Acknowledging the importance of application to reinforce knowledge and an understanding of the concepts, the students were supportive of the idea of introducing real-world examples in their classroom, as they believed that this would also make the mathematics more interesting.

The students engaged in this project were in first and second grade in a high school gifted classroom. As such, most of them had already learned quadratic function, and some had learned three-dimensional geometry. Before the project, the students only had a single direction of understanding, such as from the formula to graphs.

After the project, the students demonstrated an increased awareness of the concept of function and showed greater spatial literacy, and were able to give more descriptive explanations of their understanding. This is shown by the following feedback from the students.

S1: Before joining the program, mathematics just seemed like a collection of abstract objects. But now, I think of mathematics as a useful and combinable subject. These activities helped to give me a sense of the mathematics of 3-dimensional space, and I'm satisfied with the learning experience.

S2: While I had had the experience of drawing a graph in a situation based on a formula or conditions before, this was first time that I had to propose a formula or graph to fit objects. This was a new experience for me.

S3: I learned how to change 2-dimensional mathematical concepts into a 3D model. I want to make more diverse models.

Students showed a present and future awareness of the utilization of mathematics, and their spatial literacy was shown during the discussion.

S4: I can see how 3D visualization could be used in various fields of work, such as restoring relics and design.

S5: I have learned how to use formulas or equations in 3D designs and how to use the 3D printer. So I want to know more about constructing 3D-models.

S6: I am curious to know if there are any shapes we can't make with 3D printing.

In general, students did not experience much difficulty in performing the tasks of the project. Ultimately, they were able to reproduce their plans (Figure 9). They only needed time to handle the technologies for constructing the model of the three-dimensional surface. Interestingly, they took an active interest in 3D design for 3D printers.

S7: I want to learn more about specialized 3D design, such as irregular shapes or using mapping.

S8: I wish to explore more using Mathematica for 3D design.

S9: If I have another chance, I'd like to construct an object in my own style.

Figure 9. Final product

Through the students' responses, we found their mathematical perspectives changed and this project helped their changes.

5.2 *Going further*

Authentic learning has been shown to help connect students' classroom learning to the outside world. By linking what students are learning in mathematics classes to the real world, they better understand the problems they will be faced with when dealing with a constantly changing environment. In this chapter, we described a classroom project for gifted students that used authentic-learning-based teaching techniques to provide this link. We could find the students' mathematical learning experience through the real-world context.

The authentic teaching and learning of mathematics explicitly connects mathematical concepts, skills, and strategies to purposeful, relevant, and meaningful contexts, thereby promoting a deeper level of understanding in the classroom. The Programme for International Student Assessment (PISA) mathematics literacy test assessed students' abilities to apply their mathematical content knowledge and skills to a broad range of real-world problems (OECD, 2007). Meanwhile, one of the cognitive domains for the Trends in International Mathematics and Science Study (TIMSS) comprises reasoning skills to solve mathematical problems set in real life contexts (National Centre for Educational Statistics, 2009). In light of this, we need to rethink and to emphasize authentic learning experience in mathematics classes.

In addition, attempts have been made to develop STEAM-related intensive courses for high school students with future academic careers in Korea. In particular, in Korean STEAM education, students' emotional experiences and creativity are more strongly emphasized as important elements of their learning processes. 3D printing is able to fulfil both needs. We investigated how 3D printers could be used in the classroom, particularly for STEAM programs. This chapter shows an example of how 3D printing created benefits for a secondary school mathematics class in Korea.

When applying 3D printing technology in the mathematics classroom, there are many challenges, including the need to purchase printers, develop suitable programs, and build user skill. But working with 3D printing technology can be highly rewarding, and students love as well as need the technology. There is a great deal of activity, funding and exploration in the 3D printing space. With the drop in 3D printer prices and increase in the amount of free 3D software, it is very likely that this field will become even more accessible within the next few years. There will be more lesson plans, more interest in 3D design, and more interest in STEAM fields as this technology becomes popular and available in schools. Authentic learning through 3D printing is one method of promoting STEAM education.

Acknowledgement

The object in Figure 1 is from the Virtual Collection of Asian Masterpieces (2013), supported by NAVER, retrieved on March 7, 2014 from http://masterpieces.asemus.museum/masterpiece/detail.nhn?objectId=12016

References

Bennett, S., Agostinho, S., & Lockyer, L. (2005). Reusable learning designs in university education. In T.C. Montgomerie & J.R. Parker (Eds.), *Proceedings of the IASTED International Conference on Education and Technology* (pp.102-106). Anaheim, CA: ACTA Press.

Blum, W. (1993). Mathematical modelling in mathematics education and instruction. In T. Breiteig, I. Huntley, & G. Kaiser-Messmer (Eds.), *Teaching and learning mathematics in context* (pp. 3-14). Chichester, UK: Horwood.

Borthwick, F., Bennett, S., Lefoe, G., & E. Huber (2007). Applying authentic learning to social science: A learning design for an inter-disciplinary sociology subject, *Journal of Learning Design, 2* (1), 14-24.

Brown, J. S., Collins, A., & Duguid, P. (1989). Situated cognition and the culture of learning. *Educational Researcher, 18*(1), 32-42.

Collins, A., Brown, J. S., & Newman, S. E. (1989). Cognitive apprenticeship: Teaching the crafts of reading, writing, and mathematics. In L.B. Resnick (Ed.), *Knowing, learning and instruction: Essays in honour of Robert Glaser* (pp. 453-494). Hillsdale, NJ: LEA.

Cox, L. (2011). Manufacturers turn to 3D Printing. In D. Rotmans (Ed.) *Advanced Manufacturing, 10,* Massachusetts: Technology Review.

Crawford, G. W., & Lee, G. (2003). Agricultural origins in the Korean peninsula. *Antiquity, 77*(295), 87-95.

Dewey. J. (1897). *My pedagogic creed* (reprinted from The School Journal 54, 3, 77-80.). Carbondale, Illinois: Center for Dewey Studies.

Donovan, M. S., Bransford, J. D., & Pellegrino, J. W. (Eds.). (1999). *How people learn: Bridging research and practice.* Washington, DC: National Academy Press.

Elliot, C. (2007). Action research: Authentic learning transforms students and teacher success. *Journal of Authentic Learning, 4*(1), 34- 42.

Forman, E. A., & Cazden, C. B. (1985). Exploring Vygotskian perspectives in education: The cognitive value of peer interaction. In J.V. Wertsch (Ed.), *Culture, communication and cognition: Vygotskian perspectives* (pp. 323-347). Cambridge: Cambridge University Press.

Gardner, H. (2006). *Multiple intelligences: new horizons.* New. York: Basic Books.

Hart, G. W. (2005). Creating a mathematical museum on your desk, *Mathematical Intelligencer 27*(4), 14–17.

Herrington, J. (2006). *Authentic e-learning in higher education: Design principles for authentic learning environments and tasks.* In: World Conference on E-Learning in Corporate, Government, Healthcare, and Higher Education (ELEARN) 2006, 13-17

October 2006, Honolulu, Hawaii, USA.

Herrington, A., & Herrington, J. (2006). What is an authentic learning environment? In A. Herrington & J. Herrington (Eds.). *Authentic learning environments in higher education* (pp. 1-13). Hershey, PA: ISP.

Herrington, J., & Oliver, R. (2000). An instructional design framework for authentic learning environments. *Educational Technology Research and Development*, *48*(3), 23-48.

Herrington, J., Oliver, R., & Reeves, T. C. (2003). Patterns of engagement in authentic online learning environments. *Australian Journal of Educational Technology*, *19*(1), 59-71.

Herrington, J., Reeves, T., Oliver R., & Woo, Y. (2004). Designing authentic activities for Web-based courses. *Journal of Computing in Higher Education*, 16(1), 3-29.

Herrington, J., Specht, M., Brickell, G., & Harper, B. (2009). Supporting authentic learning contexts beyond classroom walls. In Koper, R. (Ed), *Learning network services for professional development* (pp. 273-288). Berlin Heidelberg: Springer.

Jonassen, D. H., Carr, C., & Yueh, H. P. (1998). Computers as mind tools for engaging learners in critical thinking. *Tech Trends*, *43*(2), 24–32.

Knill, O., & Slakovsky, E. (2013). *Illustrating mathematics using 3D printers*. Retrieved from: http://www.math.harvard.edu/~knill/3dprinter/documents/trieste.pdf

Kross, R. B. (2011). *How 3D printing will change absolutely everything it touches*, available at http://www.forbes.com/sites/ciocentral/2011/08/17/how-3d-printing-will-change-absolutelyeverything-it-touches/

Lombardi, M. M. (2007). Authentic learning for the 21st century: An overview. Dana G. Oblinger (Ed.) *Learning initiative-advancing learning through IT innovation*. ELI Paper 1: 2007. Educause.

Lave, J. (1988). *Cognition in practice mind mathematics and culture in Everyday Life (Learning in Doing)*. Cambridge: Cambridge University Press.

Lave, J., & Wenger, E. (1991). *Situated learning: Legitimate peripheral participation*. Cambridge: Cambridge University Press.

Meyers, N. M., & Nulty, D. D. (2009) How to use (five) curriculum design principles to align authentic learning environments, assessment, students' approaches to thinking, and learning outcomes. *Assessment and Evaluation in Higher Education*, *34*(5), 565-577.

National Centre for Educational Statistics (2009). *Highlights from TIMSS 2007: Mathematics and science achievement of U.S. Fourth and Eighth-Grade students in an international context*. Retrieved from: http://nces.ed.gov/pubs2009/2009001.pdf

New Media Consortium (2013). *NMC Horizon report- 2013 Higher Education Edition*. Retrieved from: http://www.nmc.org/pdf/2013-horizon-report-HE.pdf

OECD (2007). *PISA 2006: Science competencies for tomorrow's world*. Volume 1: Analysis. Retrieved from: http://www.pisa.oecd.org/dataoecd/30/17/39703267.pdf

Piaget, J., & Inhelder, B. (1974). *The child's construction of quantities*. London: Routledge and Kegan Paul Ltd.

Roschelle, J., & Behrend, S.   (1995). The construction of shared knowledge in collaborative problem solving. In C. O'Malley (Ed.), *Computer-supported collaborative learning* (pp. 69-97). Berlin: Springer-Verlag.

Segerman, H. (2012). 3D Printing for mathematical visualisation. *The Mathematical Intelligencer, 34,* Issue 4, 56-62.

Time (2013). *Harvard uses 3-D printing to replicate ancient statue,* available at http://style.time.com/2013/01/30/harvard-uses-3-d-printing-to-restore-ancient-statue/

Wagner, T. (2008). *The global achievement gap.* New York: Basic Books.

Wolf, A. (2013). *3D printers in the classroom: 7 Reasons why every school should have a 3D printer,* available at http://airwolf3d.com/blog/2013/02/27/school-3d-printers-in-the-classroom/

Chapter 5

What do Teachers Need to Know to Teach Secondary Mathematics?

Kim BESWICK

It is well established that teachers of mathematics need to know the content that they teach but also much more. Another aspect of teacher knowledge that has received considerable attention among mathematics educators since it was first proposed in the late nineteen eighties is pedagogical content knowledge (PCK). PCK is more than a simple combination of content and pedagogical knowledge although it depends upon both of these. Teachers use PCK to provide learning experiences that enable students to learn mathematics meaningfully. This chapter provides examples of mathematics content knowledge (MCK) and PCK items used in an online survey designed to measure these aspects of teacher knowledge. Data about the choices of small numbers of Australian and Singaporean mathematics teachers as well as those of Australian pre-service secondary mathematics teachers are provided. The main focus, however, is on ways in which the items could be used in professional learning and the kinds of knowledge teachers might need in order to use these sorts of items to create learning experiences that promote mathematics learning.

1 Introduction

Attempts to describe what teachers need to know in order to facilitate students' learning have highlighted the complexity of the knowledge that teachers require. Work over the past several decades on mathematics teachers' knowledge has built on Shulman's (1987) proposed seven categories of teacher knowledge. These were not specific to any particular subject and comprised content knowledge; pedagogical knowledge; pedagogical content knowledge (PCK); knowledge of curricula, how students learn, the contexts in which education is conducted, and the aims and purposes of education. Of these, PCK has attracted most interest among mathematics educators and has been considered in relation to that subject in a range of ways. Conceptualisations of PCK such as those of Ball, Thames and Phelps (2008) and Chick (2007) acknowledge and attempt to tease out its connections with and distinctiveness from mathematics content knowledge (MCK). Others have attempted an integrated approach to mathematics teachers' knowledge. For example, Beswick, Callingham and Watson (2012) reported a measure of a rich construct of teacher knowledge that included all seven of Shulman's knowledge types and also teachers' beliefs and confidence.

There is consensus that MCK alone is insufficient for effective mathematics teaching (Baumert et al., 2010; Hill et al., 2007; Zazkis & Leikin, 2010), nevertheless, there is agreement that MCK is necessary. It has been argued that the MCK required by teachers is different from that required by other numerate adults (Ball et al., 2008). They provided the example of knowledge of rectangle perimeter: it is reasonable to expect that an educated adult would know how to calculate the perimeter of rectangle but the knowledge a teacher needs must include an ability to, for example, think about rectangle perimeter in ways that enables him/her to analyse the mathematical quality of a student's unanticipated generalisation about rectangle perimeter. Ball et al. distinguished the MCK needed by teachers from that of other adults by calling it specialised content knowledge.

Efforts to measure teachers' knowledge of various kinds have been motivated by the assumption that more knowledgeable teachers will be more effective than less knowledgeable teachers in facilitating students' learning (Beswick, Callingham, & Watson, 2012). Over time there has been recognition that measures of teachers' knowledge need to be more subtle than considerations of the amount of the mathematics a teacher has studied. Attempts have been made to measure the kinds of knowledge that teachers need including PCK conceptualised in various ways. Those aimed at developing theoretical understanding of the nature of mathematics teachers' knowledge have tended to use intensive methods such as detailed analyses of classroom observations (Ball & Bass, 2000), interviews with teachers (Zhou, Peverly, & Xin, 2006), and assignment tasks completed by pre-service teachers (Chick & Pierce, 2008). Approaches of this sort are difficult to scale and hence there has been a major effort to develop pen and paper instruments, often comprising multiple choice items or combinations of multiple choice and short answer questions (Hill et al., 2007).

2 Mathematics Content Knowledge

Concerns about teachers' MCK and hence most research in the area, has focussed on primary school teachers (e.g., Goulding, Rowland, & Barber, 2002; Hill, 2010) and those teachers of secondary school mathematics without levels of tertiary mathematics considered appropriate (usually at least second year university level) (Vale, 2010). Major studies of teacher preparation such as the Teacher Education and Development Study in Mathematics (TEDS-M) and the Mathematics Teaching in the 21st Century (MT21) have encompasssed pre-service primary and junior secondary school teachers (Döhrmann, Kaiser & Blömeke, 2012; Schmidt et al., 2008) but not teachers of more senior year levels. Baumert et al., (2010) investigated the mathematics achievement of Year 10 classes and their teachers' MCK and PCK. MCK did not predict students' achievement whereas PCK did. The strong correlation that they found between MCK and PCK, however, suggests that PCK is dependent upon MCK. Baumert et al. (2010) also made the point that teachers require conceptual understanding of the mathematics that they teach and not simply knowledge of procedures. This explains the finding of Beswick and Callingham (under review) that, in spite of having studied at least some tertiary level mathematics, some pre-service secondary mathematics teachers in Australia had difficulty with content items that demanded a degree of analysis and the capacity to connect different mathematical ideas.

Singapore, but not Australia, participated in TEDS-M. For pre-service junior secondary school teachers Singapore ranked third of the 15 participating countries in relation to MCK behind Taiwan and Russia and ahead of other western countries that participated (Hsieh, Wong, & Wang, 2013) suggesting that Singaporean teachers have relatively high levels of MCK.

3 Pedagogical Content Knowledge

PCK has been conceptualised by mathematics educators in variety of ways that have built upon Shulman's (1987, p. 8) original description:

> "It represents the blending of content and pedagogy into an understanding of how particular topics, problems, or issues are organized, represented, and adapted to the diverse interests and abilities of learners, and presented for instruction."

Some, like Ball and colleagues (e.g. Ball et al., 2008), have subdivided PCK into specific knowledge types. Each of their categories - knowledge of content and students, knowledge of content and teaching, and knowledge of content and curriculum - emphasise the role of mathematical content knowledge in PCK. Chick and colleagues (e.g., Chick, 2007) described a spectrum of knowledge that includes aspects that are "clearly PCK" along with others in which the content or pedagogical aspects are relatively more prominent. These categories were described as content knowledge in a pedagogical context (such as when a teacher identifies the aspects of mathematics that are key to understanding a concept) and, conversely, pedagogical knowledge in a content context (such as when a teacher employs strategies for engaging students in a mathematics lesson).

Beswick and Goos (2012) reported that Australian primary pre-service teachers found PCK items more difficult than mathematics content knowledge items and Beswick and Callingham (under review) made a similar finding in relation to pre-service secondary mathematics teachers. Difficulty with agreeing on the scoring of multiple choice PCK items was acknowledged as a possible contributor to the apparent difficulty of these items (Beswick & Goos, 2012) and highlights the very contextual nature of PCK.

In TEDS-M two aspects of PCK were addressed: knowledge of curriculum and planning for mathematics teaching, and knowledge for carrying out mathematics teaching and learning activities (Döhrmann et al., 2012). In this study Singapore also ranked third to Taiwan and Russia in relation to the PCK of its pre-service junior secondary mathematics teachers (Hsieh et al., 2013).

4 The CEMENT Project

The items presented in the following sections were developed in the context of a 2-year project funded by the Australian Learning and Teaching Council and titled Building the Culture of Evidence-based Mathematics Education for New Teachers (CEMENT). The project's principal aim was to contribute towards a national culture of evidence-based practice in relation to mathematics teacher education but interest from in-service teachers and, later mathematics teacher educators lead to the instruments developed being used in other contexts.

4.1 *The survey*

Practical reasons led to the decision to develop an online automatically scored survey instrument. Item formats were thus limited to multiple choice, true/false or Likert scales. It was considered that pre-service teachers who volunteered to participate would be unlikely to spend more than an hour completing the survey which imposed limits on the numbers of items to which they could be asked to respond. The survey, therefore, comprised 10 content knowledge items, 10 PCK items, 9 beliefs items and 1 confidence item. The content knowledge and PCK items with which each participant was presented were drawn from a larger pool of items so not all items were available to all participants. This means that the numbers of respondents to particular items were less than the total numbers of participants.

Mathematical content items were developed to cover the major content strands of the Australian Curriculum: Mathematics – Number and Algebra, Measurement and Geometry, and Probability and Statistics (ACARA, 2012) as well as to include items requiring conceptual understanding, procedural understanding, and skills. PCK items were developed that fitted each of the following categories from Chick's (2007) framework: (1) analysing / anticipating / diagnosing student thinking, (2) constructing / choosing tasks / tools for teaching, (3) knowledge of representations, and (4) explaining mathematical concepts. Separate pools of both MCK and PCK items were developed for primary and secondary teachers with both groups being offered items aimed at Years 7 and 8 (the first two years of secondary school in Australia).

Many of the items, particularly those designed to measure PCK have been used in workshops with teachers and several of the universities that participated in CEMENT as well as others are using the items to stimulate discussion among pre-service teachers.

4.2 *Participants*

Pre-service secondary mathematics teachers were recruited through the seven participating universities while Australian in-service teacher participants were recruited through the website of the Australian Association of Mathematics Teachers. Singaporean teachers attending the 2013 annual conference of the Association of Mathematics Educators and the Singapore Mathematical Society were invited to complete the online survey in the early part of 2013. Most participants in the overall study were Australian primary pre-service teachers. Results for this cohort were reported by Beswick and Goos (2012). In terms of secondary teachers a total of 100 pre-service mathematics teachers, 13 Australian and 7 Singaporean in-service mathematics teachers completed the survey. It is their data that are provided along with the example items in the following sections.

5 Examples of Content Knowledge Items

The examples provided here are all from the Number and Algebra strand and selected because they had the greatest numbers of in-service teachers responding to them as well as attracting responses from more than half of the pre-service teachers. The responses are illustrative only and suggest possibilities that could be explored in a larger study. The three items are shown in Figures 1, 2 and 3. Each figure is followed by a table (Tables 1, 2 and 3) showing the numbers of Singaporean and Australian pre-service and in-service teachers who selected each of the possible answers. In each case the correct answer is marked with an asterisk.

The Tile Pattern item shown in Figure 1 was among the most difficult MCK items requiring separate consideration of odd and even side lengths and involving quadratic relationships in both cases. However, because the question asked about an even side length (50) participants only needed to consider the even case and could have observed that in such cases half of the tiles in each row are white and hence calculated the answer as $25 \times 50 = 1250$, without formally generalising.

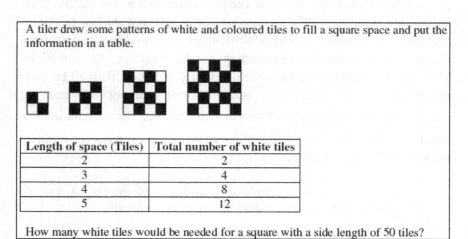

A tiler drew some patterns of white and coloured tiles to fill a square space and put the information in a table.

Length of space (Tiles)	Total number of white tiles
2	2
3	4
4	8
5	12

How many white tiles would be needed for a square with a side length of 50 tiles?

Figure 1. Tile pattern question

This item provides a useful example in which to highlight the importance of helping students to link the pictorial and numerical representations of a relationship – in this case the pictures show more clearly than the table that the odd and even cases are different. As Table 1 shows, the majority of participants selected the correct answer to this question although there is some evidence that the pre-service teachers struggled rather more than their in-service colleagues.

Table 1

Teachers' responses to the tile pattern question

Possible answer	No. Australian pre-service teachers (n=51)	No. Australian teachers (n=10)	No. Singaporean teachers (n=3)
250	4	1	1
1200	13	3	0
1250*	30	6	2
2500	4	0	0

The Flagpole Height question (Figure 2 and Table 2) presented no difficulty for any of the participants. It deals with a content area, proportional reasoning, that is of central importance to the middle years curriculum (Lamon, 2007). Discussions of anticipated student responses and ways to respond to students' difficulties so as to develop conceptual rather than procedural understanding would, therefore, be potentially very useful. In particular, the value of solution methods likely to be more meaningful to at least some students than solving the standard proportion, $\frac{100}{60} = \frac{x}{5.4}$, would be worth discussing. These could include less formal consideration of equivalent ratios or representing the problem using a double number line.

An upright 1-metre stick casts a shadow that is 60 cm long. At the same time, a flagpole casts a shadow that is 5.4 metres long.

How high is the flagpole?

Figure 2. Flagpole height question

Table 2

Teachers' responses to the flagpole height question

Possible answer	No. Australian pre-service teachers (n=62)	No. Australian teachers (n=10)	No. Singaporean teachers (n=3)
5.8 m	0	0	0
6 m	0	0	0
9 m*	62	10	3
10.8 m	0	0	0

The next item, Think of a Number, shown in Figure 3, involved translating a description of a series of operations on an unknown number into an algebraic expression. Table 3 shows that only a very small number of participants chose an incorrect answer. The single teacher who answered this question incorrectly appears to have confused squaring with multiplying by 2. Activities of this sort are useful ways to assist students to appreciate the role of the pronumeral as an unknown number. Building expressions up in this way can help to develop understandings that later will allow equations to be solved.

In contrast with this item, illustrative of a symbolic orientation to algebra, the Tile Pattern item exemplifies a patterns, or function approach to algebra. Both approaches, and combinations of them, can be found in contemporary curricula but there is a lack of research evidence as to which of these approaches is preferable (Kieran, 2007; Stacey & MacGregor, 2001). The greater apparent difficulty of the Tile Pattern item for participants in this study is consistent with Even's (1993) report of pre-service secondary mathematics teachers' limited conceptual understanding of functions. The pair of items could stimulate worthwhile discussions of the differing conceptual underpinnings and relative affordances of the two approaches.

> Claire thinks of a number, n.
> She multiplies the number by itself.
> She then halves that answer and subtracts 10.
> Which expression shows what Claire did?

Figure 3. Think of a number question

Table 3

Teachers' responses to the think of a number question

Possible answer	No. Australian pre-service teachers (n=61)	No. Australian teachers (n=10)	No. Singaporean teachers (n=3)
$\dfrac{2n-10}{2}$	0	0	0
$\dfrac{2n}{2}-10$	2	1	0
$\dfrac{n^2}{2}-10$ *	59	9	3
$\dfrac{n^2-10}{2}$	0	0	0

6 Examples of Pedagogical Content Knowledge Items

Figures 4, 5, 6 and 7 show four of the items designed to measure teachers' PCK. Each figure is followed by a table (Tables 4, 5, 6 and 7) showing each of the possible answers and the numbers of Singaporean and Australian teachers who selected each. The preferred answer for each example is marked with an asterisk.

The first example, Mixing cordial, is shown in Figure 4. It was aimed at assessing participants' ability to diagnose student thinking. As shown in Table 4, two Australian teachers responded incorrectly to this question along with more than one quarter of the pre-service teachers. Although designed to measure PCK, answering correctly relies on knowing that to maintain the sweetness of a cordial mix the concentrate and water need to be kept 'in proportion'. That is, there is a multiplicative relationship between the quantities of concentrate and water in equally sweet cordial.

The following question was given to Year 8 students:

Some children are making batches of cordial by mixing together sweet concentrate and water.

Sally uses 4 cups of sweet cordial mix and 13 cups of water.

Myles uses 6 cups of sweet cordial mix and 15 cups of water.

One student thinks these cordial mixes will have the same sweetness.

Which of the following cordial mixes might he ALSO think is the same sweetness?

Figure 4. Mixing cordial question

Table 4

Teachers' responses to the mixing cordial question

Possible answer	No. Australian pre-service teachers (n=53)	No. Australian teachers (n=9)	No. Singaporean teachers (n=3)
Aisha uses 8 cups of sweet cordial mix and 26 cups of water.	7	1	0
Carly uses 10 cups of sweet cordial mix and 19 cups of water.*	37	7	3
Deng uses 8 cups of sweet cordial mix and 20 cups of water.	4	0	0
Erin uses 10 cups of sweet cordial mix and 28 cups of water.	5	1	0

Participants who selected the first option, which describes a mix that would be the same sweetness as Sally's, may not have appreciated that the children who thought Sally and Myles had made equally sweet mixes are likely to have focused on the fact that Myles had two more cups of concentrate and two more cups of water that Sally – an additive rather than multiplicative relationship. The third option has no relationship, either additive or multiplicative, with Sally's or Myle's cordial.

Choosing it may reflect a belief that students' incorrect responses are haphazard rather than based on systematically applied misconceptions as is more usually the case. The numbers in the final option were obtained by adding the concentrate and water components of Sally's and Myle's cordial. It is the kind of response that some students make when they see numbers but have little idea of the meaning of the question and hence how to solve the problem. Choosing the 'correct' option was taken to be indicative of the teachers' ability to diagnose the students' thinking and to recognise a mathematically similar solution – in this case one that had an additive relationship with the components of Sally's cordial – and to recognise that students with misconceptions are likely to apply the same thinking across multiple situations.

The next problem, Function cards, is shown in Figure 5 with response frequencies in Table 5. It was designed to address the PCK required to select appropriate examples i.e. to construct or choose tasks/tools for teaching. Again the dependence of PCK on MCK is evident: participants need to know what a function is. PCK requires going beyond this to recognise the value of providing students with multiple examples and, crucially, non-examples in assisting them to develop robust concepts. Watson and Mason (2005) acknowledged controversy around the idea of providing students with non-examples but pointed out that non-examples "demonstrate the boundaries or necessary conditions of a concept" (p. 65). They can draw students' attention to defining features of classes of mathematical objects such as, in this case, functions.

Of the options, shown in Table 5, the first and last are functions whereas the second and third are not, provided in the case of the second, the domain is taken to be the set of real numbers indicated by the x-axis. The ambiguity in relation to second option meant that the 'best' answer was taken to be the third. The popularity of the fourth option suggests that, assuming they knew what a function is (by no means guaranteed), many participants and especially pre-service teachers, considered an additional example, rather than a non-example, to be the best option.

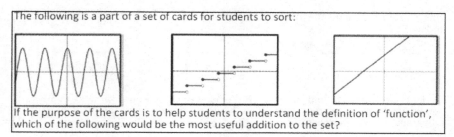

Figure 5. Function cards question

Table 5

Teachers' responses to the function cards question

Possible answer	No. Australian pre-service teachers (n=51)	No. Australian teachers (n=8)	No. Singaporean teachers (n=3)
	6	0	0
	7	0	1
	8	5	1
	30	4	1

The third PCK example, Comparing algebraic representations, was designed to assess participants' knowledge of representations of mathematical relationships. It is shown in Figure 6 and the numbers of participants choosing each of the options provided are shown in Table 6.

This diagram represents the number of people that can be seated as small tables are added.

Students were asked to find a rule to link the number of tables used and the number of people who could be seated, and express this algebraically using T for the number of tables and P for the number of people.

Three students' answers were:　　　P = 2T + 2　　　*P = (T – 2) x 2 + 6*　　P = 2(T + 1)

What representation could you best use to convince them that these solutions are the same?

Figure 6. Comparing algebraic representations question

Table 6

Teachers' responses to the comparing algebraic representations question

Possible answer	No. Australian pre-service teachers (n=53)	No. Australian teachers (n=9)	No. Singaporean teachers (n=3)
Work through the algebra on the board	17	2	2
Create a table of values for each rule and compare them*	27	6	1
Draw a graph for each rule on the same axes	7	0	0
Get them to try a different problem with the same relationship	2	1	0

Most Australian teachers and pre-service teachers agreed with the preferred option. Drawing a graph to represent each of the three equations presented in the question would require finding a series of points, and so encompasses option 2, and working through the algebra, the next most popular option for teachers overall, has merit if the students' prior learning is appropriate. Indeed a task like this could be used to create a context in which demonstrate how manipulating algebraic expressions can show their equivalence. The numbers of responses are far too small to draw conclusions about possible differences between Singaporean and Australian teachers but if there is a difference between the response patterns of the two groups it may reflect differing emphases in curricula as well as practice on less formal approaches (Australia) compared to more formal approaches (Singapore).

The final example, shown in Figure 7, relates primarily to the PCK required to explain mathematical concepts, specifically comparing decimals. As shown by the response frequencies in Table 7, there was no consensus among participants regarding the best answer to this question. All four options attracted responses from Australian teachers and pre-service teachers and the responses of the three Singaporean teachers included two options.

> Which of the following explanations is least likely to be helpful in assisting a student who is struggling to understand that 1.26 is greater than 1.026?

Figure 7. Representing decimals question

The most popular non-preferred option involves equalising the lengths of the pair of decimals to facilitate their comparison. Roche and Clarke (2006) found that this approach reliably led to correct answers but they expressed concerns about its implications for the development of conceptual understanding. Adding zeroes to make the decimals the same length 'works' because it allows students to consider the decimal parts of the numbers as whole numbers. It does not require thinking about the place value of the digits involved. Place value thinking, however, requires students to understand the relative sizes of digits in different places. Importantly, place value thinking leads to correct answers just as reliably as the whole number thinking that is encouraged by equalising lengths (Roche & Clarke, 2006). It should be noted that the although the second option uses place value language it does so in a way that does not clearly express an understanding of the place value words. In fact the statements resemble exercises found in some Australian textbooks that encourage predictable repetition rather than reasoning. In contrast to the statements in option 2 Roche and Clarke (2006, pp. 428-429) provided the following examples of place value reasoning that support both correct answers and conceptual development:

1.46 is greater than 1.45 because 1.46 is one hundredth more

0.567 is greater than 0.3 because five tenths is greater than three tenths,
or
0.567 is more than one half but 0.3 is less than a half

0.87 is greater than 0.087 because 87 hundredths is greater than 87 Thousandths

0.7 is the same as 0.70 because 7 tenths equals 7 tenths (or 70 hundredths)

Table 7

Teachers' responses to the representing decimals question

Possible answer	No. Australian pre-service teachers (n=53)	No. Australian teachers (n=9)	No. Singaporean teachers (n=3)
Multiplying both numbers by 1000 gives us 1260 and 1026. 1260 is greater than 1026 so 1.26 is greater than 1.026	9	2	0
1.26 is 1 whole + 2 tenths + 6 hundredths. 1026 is 1 whole + 0 tenths + 2 hundredths + 6 thousandths	10	2	0
It's easier to compare decimals that are the same length and we can add zeros to the end of a decimal without changing it, so 1.26 is the same as 1.260. We can now see that 1.26 is greater than 1.026 because 260 is greater than 026 (which is 26)	17	3	2
	17	2	1

The first option shown in Table 7, like the third, also promotes whole number rather than place value thinking and so is not helpful for conceptual development. This leaves the last option showing the use of decimal squares as the preferred option. There are other representations of decimals that may be better still. For example, Stacey et al. (2001) recommended a length model such as Linear Arithmetic Blocks comprising lengths of tube (washers for the smallest length) representing thousandths, hundredths, tenths and ones.

7 Using the Items in Professional Learning

MCK items, such as those presented in Section 5 of this chapter can stimulate interesting discussion among teachers and, if used sensitively can provide opportunities for teachers to deepen their conceptual understanding of mathematics content. Questions such as the following can provide useful starting points.

1. How do you anticipate students in your class would respond to this item? How would you respond?
2. Would you use this item in your teaching? With which students? For what purpose? Under what circumstances would you use/not use it? Why/why not?
3. How could you help students who have difficulty accessing the task to make a start? How could you extend the task in meaningful ways?

To an even greater extent PCK items based on teaching scenarios provide effective ways to start discussion about teaching, the knowledge that teachers draw upon as they teach, and the reasoning in which they are constantly engaged as they plan for, conduct, and reflect upon teaching. As was evident, from the examples provided in this chapter, 'correct' answers are contentious. It is possible to imagine a situation in which almost every option might be reasonable. It is because of this that these items are so useful to help teachers make their knowledge explicit.

Of possibly even greater value as a professional learning activity than discussing examples is the work of devising examples. Coming up with scenarios that draw on particular aspects of PCK for mathematics teaching along with plausible alternatives for a multiple choice format is a challenging task that reveals much about teachers' PCK. In a group context the conversations that ensue provide a rich environment for its development.

8 Conclusion

As stated already the data presented here are not sufficient to make any comparative judgments. In addition, the Australian teachers were recruited through the website of the Australian Association of Mathematics Teachers and are therefore drawn from a subset of Australian mathematics teachers who identify as teachers of mathematics to the extent that they value membership of the association.

The examples presented illustrate the usefulness of such items, especially when used in the context of collegial discussion, to uncover and enhance, mathematics teachers' knowledge.

Consistent with Baumert et al.'s (2010) finding that although PCK predicts student achievement it is underpinned by sound MCK. The PCK examples provided here all require MCK as a starting point. They also show that much more than MCK is needed for teachers to design and implement experiences that promote their students' mathematics learning.

Acknowledgement

This research was supported by the Australian Learning and Teaching Council Priority Projects grant number PP10-1638. The contributions to the development and implementation of the study of Rosemary Callingham, Helen Chick, Julie Clark, Merrilyn Goos, Barry Kissane, Pep Serow, and Steve Thornton are also acknowledged.

References

Australian Curriculum, Assessment and Reporting Authority (ACARA). (2012). *Australian curriculum: Mathematics.* Retrieved March 7, 2014, from http://www.australiancurriculum.edu.au/Mathematics

Ball, D. L., & Bass, H. (2000). Interweaving content and pedagogy in teaching and learning to teach: Knowing and using mathematics. In J. Boaler (Ed.), *Multiple perspectives on mathematics teaching and learning* (pp. 83-104). Westport, CT: Ablex.

Ball, D. L., Thames, M. H., & Phelps, G. (2008). Content knowledge for teaching: What makes it so special? *Journal of Teacher Education, 59*(5), 389-407.

Baumert, J., Kunter, M., Blum, W., Brunner, M., Voss, T., Jordan, A., & Tsai, Y.-M. (2010). Teachers' mathematical knowledge, cognitive activation in the classroom, and student progress. *American Educational Research Journal, 47*(1), 133-180.

Beswick, K., & Callingham, R. (under review). Pre-service secondary mathematics teachers' knowledge: Establishing an evidence base.

Beswick, K., Callingham, R., & Watson, J. M. (2012). The nature and development of middle school mathematics teachers' knowledge. *Journal of Mathematics Teacher Education, 15*(2), 131-157. doi: 10.1007/s10857-011-9177-9

Beswick, K., & Goos, M. (2012). Measuring pre-service primary teachers' knowledge for teaching mathematics. *Mathematics Teacher Education and Development, 14*(2), 70-90.

Chick, H. (2007). Teaching and learning by example. In J. M. Watson & K. Beswick (Eds.), *Mathematics: Essential research, essential practice: Proceedings of the 30th annual conference of the Mathematics Education Research Group of Australasia* (Vol. 1, pp. 3-21). Sydney: MERGA.

Chick, H., & Pierce, R. (2008). Issues associated with using examples in teaching statistics. In O. Figueras, J. L. Cortina, S. Alatorre, T. Rojano & A. Sepúlveda (Eds.), *Proceedings of the joint meeting of PME32 and PME-NA XXX* (Vol. 2, pp. 321-328). Mexico: Cinstav-UMSNH.

Döhrmann, M., Kaiser, G., & Blömeke, S. (2012). The conceptualisation of mathematics competencies in the international teacher education study TEDS-M. *ZDM Mathematics Education, 44*, 325-340.

Even, R. (1993). Subject-matter knowledge and pedagogical content knowledge: Prospective secondary teachers and the function concept. *Journal for Research in Mathematics Education, 24*(2), 94-116.

Goulding, M., Rowland, T., & Barber, P. (2002). Does it matter? Primary teacher trainees' subject knowledge in mathematics. *British Educational Research Journal, 28*(5), 659-704.

Hill, H. C. (2010). The nature and predictors of elementary teachers' mathematical knowledge for teaching. *Journal for Research in Mathematics Education, 41*(5), 513-545.

Hill, H. C., Sleep, L., Lewis, J. M., & Ball, D. L. (2007). Assessing teachers' mathematical knowledge: What knowledge matters and what evidence counts? In F. K. Lester Jr. (Ed.), *Second handbook of research on mathematics teaching and learning* (pp. 111-155). Charlotte, NC: Information Age Publishing.

Hsieh, F. J., Wong, K. Y., & Wang, T.-Y. (2013). Are Taiwanese and Singaporean future teachers similar in their mathematics-related teaching competencies? *International Journal of Science and Mathematics Education, 11*(4), 819-846.

Kieran, C. (2007). Learning and teaching algebra at the middle school through college levels. In F. K. Lester Jr. (Ed.), *Second handbook of research on mathematics teaching and learning* (Vol. 2, pp. 707-762). Charlotte, NC: Information Age Publishing.

Lamon, S. J. (2007). Rational numbers and proportional reasoning: Toward a theoretical framework for research. In F. K. Lester Jr. (Ed.), *Second handbook of research on mathematics teaching and learning* (Vol. 1, pp. 629-667). Charlotte, NC: Information Age Publishing.

Roche, A., & Clarke, D. (2006). When successful comparison of decimals doesn't tell the full story. In J. Novotná, H. Moraová, M. Krátká & N. Stehliková (Eds.), *Proceedings of the 30th conference of the International Group for the Psychology of Mathematics* (Vol. 4, pp. 425-432). Prague: PME.

Schmidt, W. H., Houang, R. T., Cogan, L., Blömeke, S., Tatto, M. T., Hsieh, F. J., & Paine, L. (2008). Opportunity to learn in the preparation of mathematics teachers: Its structure and how it varies across six countries. *ZDM Mathematics Education, 40*, 735-747.

Shulman, L. S. (1987). Knowledge and teaching: Foundations of the new reform. *Harvard Educational Review, 57*(1), 1-22.

Stacey, K., & MacGregor, M. (2001). Curriculum reform and approaches to algebra. In R. Sutherland, T. Rojano, A. Bell & R. Lins (Eds.), *Perspectives on school algebra* (pp. 141-153). Dordrecht: Kluwer.

Stacey, K., Helme, S., Archer, S., & Condon, C. (2001). The effect of epistemic fidelity and accessibility on teaching with physical materials: A comparison of two models for teaching decimal numeration. *Educational Studies in Mathematics, 47*(2), 199-221.

Vale, C. (2010). Supporting "out of field" teachers of secondary mathematics. *Australian Mathematics Teacher, 66*(1), 17-24.

Watson, A., & Mason, J. (2005). *Mathematics as a constructive activity*. Mahwah, NJ: Lawrence Erlbaum.

Zazkis, R., & Leikin, R. (2010). Advanced mathematical knowledge in teaching practice: Perceptions of secondary mathematics teachers. *Mathematical Thinking and Learning, 12*(4), 263-281.

Zhou, Z., Peverly, S. T., & Xin, T. (2006). Knowing and teaching fractions: A cross-cultural study of American and Chinese mathematics teachers. *Contemporary Educational Psychology, 31*, 438-457.

Chapter 6

Defining, Extending, and Creating: Key Experiences in Mathematics

Yoshinori SHIMIZU

Definitions play critical roles in understanding a concept, in reasoning, in problem solving, and in communicating with others in mathematics. Moreover, key experiences in learning mathematics include examining a given definition of a mathematical term at the phase of extending the meaning of a concept and creating a new definition for the concept in an expanded system or theory. However, students seem to have difficulties in using definitions in various contexts of learning mathematics, while they have few experience of extending the meaning of a mathematical concept on their own. In this chapter the author discusses the role and nature of definitions in teaching and learning mathematics and illustrates rich experiences provided in the process of defining, extending, and creating mathematics concepts. Data from a teaching experiment are presented to exemplify how pairs of tenth grade students could examine and create definitions of various mathematical objects, including unfamiliar quadrilaterals such as a kite and "boomerang".

1 Introduction

In mathematics a definition is simply an agreement that makes clear the meaning attached to a certain word or symbol. It plays critical roles in understanding a concept, in reasoning, in problem solving, and in communicating with others. In other words, doing mathematics cannot be possible without definitions. Moreover, key experiences in mathematics include students' activity of examining a given definition of a mathematical term when they try to extend the meaning of a concept and to create a new concept for going further in an expanded mathematical system while solving new problems.

However, many students seem to have difficulties in using definitions, explicitly or implicitly, in various phases of learning mathematics. The difficulties can be a discrepancy between concept image and concept definition (Vinner & Dreyfus, 1989; Vinner, 1991). Also, not only can students' examples be inconsistent with conventional mathematical definitions, but their examples are sometimes inconsistent with their own definitions (Wilson, 1986; Zazkis & Leikin, 2008). Another students' difficulty in understanding and using definitions lies in appreciating their important role in mathematical argumentation or proof. These difficulties may be derived from the shortages of students' experience of examining given definitions from various perspectives and creating their own definitions in doing mathematics.

In the Japanese context, the meaning and role of definitions in mathematical reasoning are introduced to the students in eighth grade, when they learn the notion of proof in plane geometry (Ministry of Education, Culture, Sports, Science, and Technology, 2009). In most textbooks, the meaning of "definition" is explained by using a familiar concept like an isosceles triangle as an example. It should be noted that it is one thing for students to know the definition of a concept, an isosceles triangle, for example, and that it is quite another for them to appreciate the role of definitions in mathematics. Many, if not most, students seem not to feel the need for a definition even when they are using it. Students have few opportunities for creating a definition of a mathematical concept and they are often unaware of both the constructive and tentative nature of definitions in mathematical reasoning and the important role in communicating with others.

In this chapter the author discusses the role and nature of definitions in teaching and learning mathematics and illustrates rich experiences provided in the process of defining, extending, and creating mathematics concepts. The data from a teaching experiment (Shimizu, 1997) is used to exemplify how pairs of tenth grade students could examine and create definitions of various mathematical objects, including unfamiliar quadrilaterals such as a kite and "boomerang".

2 Examining and Creating Definitions in Doing Mathematics

The importance of definitions in doing mathematics is crucial. Definitions play critical roles at various phases of doing mathematics; understanding and using mathematical concepts and symbols, reasoning and problem solving in various areas in mathematics, and communicating with others in general, and proof and proving, in particular.

Defining is a key to creative mathematical activity. It should be noted that making a definition in doing mathematics is not a straightforward process. It involves various phases of activities, not only using and creating concept images and concept definitions but also formulating, negotiating, and revising a definition (Zandieh & Rasmussen, 2010). Key learning experiences then involve such activities as creating tentative definition for the purpose of problem solving, generating conjectures in mathematical inquiry with providing and examining examples fit to a definition, creating a new definition for extending the meaning of mathematical concepts in a broader context.

A look at the history of mathematics tells us the importance of communication with others for using a definition in the context of proof. Proof in nature is a technique for persuading others as well as oneself. Historically, the discrimination among the terms axiom, definition, and postulate became clearer when "reflections" by the people on the theory became deeper and sharper through the process of using and revising such a technique (Nakamura, 1971). This aspect of proof suggests the importance of social process for recognizing the need for a definition.

Any term in mathematics can be found to have different definitions even in textbooks of mathematics. The author does not have to select the same definition that others have used. For example, a trapezoid is defined in two different ways (Popovic, 2012); (a) A trapezoid is a quadrilateral that has *at least one* pair of opposite sides, which are parallel, and (b) A trapezoid is a quadrilateral that has *exactly one* pair of opposite sides, which are parallel. These two definitions determine at least two different hierarchical classifications of quadrilaterals. Then, examining two definitions can provide learning opportunities for understanding the arbitrary nature of definitions. As for kites, Usiskin and Griffin (2008) found that there were six different definitions among 18 textbooks they investigated in the United States. For example, a kite can be defined as "a quadrilateral which has two distinct pairs of adjacent sides of the same length" or "a quadrilateral in which exactly one diagonal is a perpendicular bisector of the other." The properties or attributes of the object focused in the definition can be the sides of equal length or symmetry with a perpendicular bisector.

Comparing the different definitions and thinking about the implications of having different definitions raise an opportunity for the students to see that they can make different hierarchical classifications, depending on the specific definition that is in use. However, students are not aware that there is a choice of definitions for mathematical terms and that they could choose a definition of a term from their standpoint.

In a classical example described in *"The Nature of Proof"* (Fawcett, 1938/1995), that aimed to foster students' "critical thinking", definitions and propositions were socially constructed by students and the teacher. More recently, Borasi (1992, 1994) conducted a teaching experiment in which the students experienced the need for monitoring and justifying their mathematical work when they were engaged in "error activities", which included activities such as defining the familiar notion of a circle or creating a definition for an unfamiliar concept of polygon. These studies suggest the importance of students' experience in writing and discussing their own definitions and then using them. Also, these studies suggest the significance of the roles of others in a social construction of a mathematical definition in appreciating its nature and roles.

It is important to notice that defining a concept in mathematics is done in a certain way when the concept is extended into a broader context. Extension is a key activity in mathematics and definition plays a crucial role. For example, the meaning of exponent "n" in a^n is defined as repeated multiplication of "a", where "a" is a positive number, at the beginning. However, to maintain the law of exponent in a larger set, the meaning of exponent needs to be extended. As for "$a^m \div a^n = a^{(m-n)}$" for natural numbers "m" and "n" (m>n), for instance, can be maintained for both m = n and m < n. Thus, a^0 is defined as "1" so that the law of exponent can be maintained.

3 Defining, Extending, and Creating: An Example

In this section, the data from a teaching experiment is presented to exemplify how pairs of tenth grade students could examine and create definitions of various mathematical objects, including unfamiliar quadrilaterals such as a kite and "boomerang" and raised a new definition of exterior angle to an obtuse angle when they tried to find the sums of exterior angles of these figures.

3.1 *Students*

A teaching experiment including sixteen instructional experiences over four months was conducted with two pairs of tenth grade students (8 experiences with each pair) in a high school in metropolitan Tokyo. The students were selected based on their responses to a preliminary questionnaire survey which had been conducted with 35 tenth grade students one month before the teaching experiment to identify their backgrounds. The items of the questionnaire included broader questions about the meaning of the term "definition" and "theorem" respectively (e.g. "Describe briefly the meaning of the word definition.") , as well as more specific questions about "kite" (e.g. "Give the most concise name for the figure", asking the name of kite by showing a figure of it.).

Four students, students F & I as Pair I (males), and students S & U as Pair II (females) selected for the teaching experiment were those who were familiar with kite but who could not differentiate the meaning of "definition" from that of "theorem". They have already studied proof in the previous grades. They were regarded as "above the average" or "average" students by their mathematics teacher, based on their achievements on regular tests in their school.

3.2 *Procedure*

The teaching experiment was designed and taught by the author with a particular format for implementing. A series of tasks was given to each pair of the students that asked them in pairs to find some properties and to think of definitions of various quadrilaterals including kite and "boomerang". Those properties and definitions were organized in such a way that they would appear necessary for the solution to certain problems. Students' problem-solving activities by themselves were supposed to last about 30 to 40 minutes, followed by interview/instructional sessions of about 20 minutes in duration. Basically, there was no intervention by the teacher (researcher) during students' problem solving, the first phase of each session. All the activities in the teaching experiment were audiotaped and videotaped. Transcribed protocols were made and submitted to analysis.

3.3 *Teacher's roles*

The teacher began each interview session by asking the students to explain their entire solution process by using their written work immediately after solving a problem. When the students' explanation was not clear to him, the teacher requested a further explanation. Then he asked the reason why they had thought in a certain way at the key points of their solution process. In case the students could not solve the problem by themselves, the teacher gave a suggestion and then explored the solution with them. The teacher's role in the interview/instructional sessions, was supposed to include: (1) making the students' solution process explicit by having them reflect on it, (2) asking them to justify their solutions, and (3) facilitating their deeper understanding of the properties and definitions of quadrilaterals through a discussion with them. In addition, the teacher made a conscious effort to ask "why questions" at certain points in each session in order to draw a well-grounded explanation from the students.

3.4 *Tasks*

The tasks for the sessions were selected with a focus on the properties and definitions of kite and "boomerang" for several reasons. First, these figures were supposed to be familiar to the students, but not taught within the framework of Japanese national curriculum guidelines. This condition was considered so that each pair of students would start on an equal footing. Second, because these figures certainly have nice properties to think about and hence have potential for producing learning opportunities, the students were supposed to be really involved in a series of tasks. Third, the students have already learned several characteristics of other related figures like parallelogram and rhombus in the previous grade. Therefore opportunities for reflecting on their own knowledge about those related figures were supposed to take place when they were working on the tasks.

A series of sessions focusing on the definitions and properties of kite and boomerang was planned, and then implemented with minor revisions in response to the students' progress. The topics for the sessions are described in Table 1.

Table 1

Tasks for sessions

Session	Task
Session 1	Classifying and naming unfamiliar quadrilaterals
Session 2	Classifying quadrilaterals from a viewpoint of "equal length"
Session 3	Properties of a kite and a formula for finding its area
Session 4	Is a boomerang a kite? Defining a kite
Session 5	Defining familiar figures
Session 6	Classifying and defining a "riple"*
Session 7	Examining definitions of kite proposed by the other students
Session 8	Finding sums of interior and exterior angles of a kite and a boomerang

*A "riple" is a quadrilateral with two obtuse angles. In this session, students are asked to identify "riples" among various figures presented in the worksheet.

In session 8, in particular, on which this chapter focuses, students were working on the task of finding the sums of the interior and exterior angles of both a kite and a boomerang.

Find the sums of the interior and exterior angles respectively of the kite and the boomerang. An exterior angle (to the <BAC) is shown in figure below.

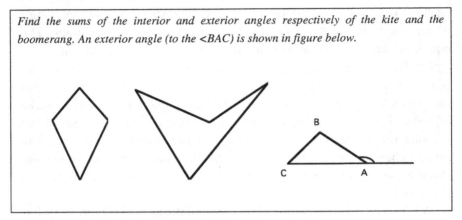

Figure 1. A task for session 8

The task was supposed to raise a new problem for the students to think about during their problem-solving activity. Namely, the task was given to the students with the intention of asking them to consider what definition of exterior angle might be appropriate for the "extended" case, since the ordinary definition of it (Figure 3) does not seem to fit in with the one for the obtuse angle in "boomerang" (Figure 2).

3.5 *Results*

Data sources
The date sources consisted of the students' responses to the preliminary questionnaire, transcribed protocols of both problem solving and interview/instruction sessions, and students' written work. Also, field notes by the researcher were used for making a narrative summary of each session. Special attention was given to those utterances by the students during the paired problem solving which could be regarded as indicating an influence of "others" on their problem solving. The protocols were coded in the following ways. The utterances of the

following three types were coded by the researcher as an indication of thinking "others": (a) a student was referring to another person as the subject in a sentence (e.g., *"Sensei* (the *teacher) must think this as a right angle, when he looks at it quickly."* Student F, in session 6); (b) a student was referring to a third person's behaviours (e.g., *"But, we must be in trouble if we would be referred as totally incorrect."* Student I, in session 6); (c) a student was saying something in the form of "talking" to someone else (e.g., *Let's say something like "we have tentatively done"*. Student F, in session 8). The protocols were also coded in terms of the "levels of their discourse" at which the students were examining a definition by using the following categories for the topic of their discourse: Level 0, Properties of certain figures; Level 1, A definition of a figure; Level 2, The nature of definitions.

An overview of sessions

The protocols from sessions 1, 3, 6, and 8 were analyzed to explore the extent to which the students could be reflective about their own mathematical activities and the extent to which they could be aware of the nature of definitions. Students' problem-solving activities in these sessions continued about 40 minutes on average for Pair I and about 36 minutes for Pair II, followed by the interview sessions of various time durations. Table 2 shows the elapsed time by each pair in sessions 1, 3, 6, and 8.

Table 2

Elapsed time in each session (min/sec)

	Pair			
	I (F & I)		II (S & U)	
Types of session	PS	I/I	PS	I/I
Session 1	44'32	02'38	47'35	07'30
Session 3	32'14	13'54	19'19	22'24
Session 6	47'28	24'37	37'57	08'24
Session 8	37'28	26'15	37'50	20'16
Average	40'26	16'51	35'40	14'39

PS: problem-solving activities; I/I: interview/instruction sessions

In each session students were thinking and communicating about the task at various levels of discourse. Table 3 shows the levels of discourse during students' problem-solving activities in each session. In the earlier sessions (1 & 3) the students tended to communicate at level 0, and then they had discourse at level 1 in sessions 6 & 8, and at level 2 in sessions 6 (pair I & II) and 8 (pair I).

Table 3

Levels of discourse in each session

	Pair							
	I (F & I)				II (S & U)			
Sessions	1	3	6	8	1	3	6	8
Level 0	*	*	*	*	*	*	*	*
Level 1			*	*			*	*
Level 2			*	*			*	

Note: A star in each cell represents that students' discourse was observed at this level.

The analysis of the "levels of discourse" suggest that in order for their concept of definition and its role in a formal system to be expanded, students need to engage in the activities across various levels of discourse.

3.6 *An overview of students' problem solving: Session 8 with Pair II*

Students' problem-solving activities in session 8 by pair II lasted about 38 minutes, followed by the interview/instructional session of about 20 minutes duration. Students' activity on examining the definition of an exterior angle, which was supposed to be a focus in session 8, is briefly summarized below.

After having concluded that the sums of interior angles of both kite and boomerang should be 360°, they found that the sum of exterior angles of kite is also 360° by applying the relationship between an exterior angle and its interior opposite angles in a triangle. Then they started to discuss the case of boomerang.

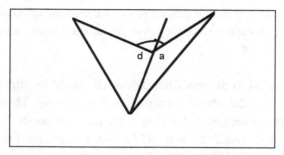

Figure 2. Dividing the obtuse angle into two parts

When they found the sum of exterior angles of the kite, the students used the definition of an exterior angle as the angle between the line and an extension of the other line at a point. They tried to apply this definition to the case of boomerang. Dividing the obtuse angle into two parts, namely, "a" and "d" in Figure 2, student U proposed that the exterior angle to the obtuse angle might be "360° - (a+d)". Responding to this idea, student S pointed out that this idea could be inconsistent with the definition of an exterior angle they had used so far, by showing another exterior angle that seemed "too big" (Figure 3).

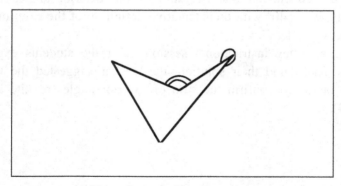

Figure 3. A new idea of exterior angle

Student S: why, however,......if the exterior angle....might be here, we should do the same thing to other angles.
Student U: exterior angles to other angles?

Student S: Yahh, .. 'cause, it doesn't fit in with the case of other angles so far. Then, 360 minus this angle, about 30, makes 330. It's why this guy is too big, isn't it ?

They then started to discuss what definition would be appropriate for the exterior angle to the obtuse angle of the boomerang. They were getting stuck and often mentioned that they were not sure about the definition of an exterior angle. (e.g., *"First of all I haven't understood the definition of an exterior angle well."* Student S).

After having spent about 9 minutes for the discussion, they finally concluded that the exterior angle for the obtuse angle was the "opposite angle", the angle gotten by subtracting the obtuse angle from 360°. Their basic idea about the definition of an exterior angle was that *"an exterior angle should be located outside to the figure"* (Student U). They found that the sum of exterior angles of boomerang was 540° by applying the definition that student U had proposed, and that by using the definition they would lose the "consistency" with the original one.

In summary, they found the sums of interior angles of both the kite and the boomerang and the sum of the exterior angles for the kite, but reached the conclusion that the sum of the exterior angles for the boomerang was 540° with their tentative definition of the exterior angle of the obtuse angle.

In the interview/instructional session, after the students explained how they had found their solution, the teacher suggested the way of "going around" to confirm the sum of exterior angles of kite is 360° (Figure 4).

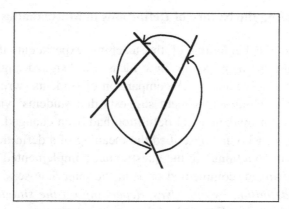

Figure 4. "Going around" exterior angles

The teacher tried to make sure about their definition of an exterior angle, asking *"What would be your answer, if you were asked what the definition is for an exterior angle?"* Student S responded as *"the angle gotten by subtracting it from 180°."* The focus of discussion was on how they could get an exterior angle by drawing an extension of one of the two lines that constitute the angle. Then he asked the students to examine the case of the boomerang, asking how they could "subtract" the obtuse angle from 180° in the same way as the case of a kite. Student S immediately responded *"we can't do it."* Student U then followed with *"We may have a negative something."* The students naturally proposed the idea of a "negative angle".

Then the teacher asked the students *"What happens to the sum of the exterior angles of the boomerang, if we might think of this guy as a negative?"* They finally concluded that they could keep the sum of the exterior angles for the boomerang as 360° by introducing the idea of a negative angle. Then, during the conversation between the students, student U proposed her idea of "signed angles" to incorporate two cases. At this point of the session, they were very flexible both in proposing and accepting new ideas concerning an extended definition of exterior angle. The teacher then suggested the idea of a directed angle to incorporate the two cases.

4 Understanding the Nature of Definitions in Mathematics

In contrast to the beginning of the teaching experiment, the students showed flexibility in proposing new ideas like "signed angles" and in accepting a "negative angle". A comparison of students' writings before and after the teaching experiment suggests that students' views on the nature and role of mathematical definition had been changed. Student U, on the one hand, who had described the meaning of a definition as "what is given to explain a thing" in the questionnaire implemented prior to the teaching experiment, commented on it in the interview session that *"we can make a definition by ourselves, depending on the situation..."*. On the other hand, student S, who had described a definition as *"Since many people have gotten the same answer, it is viewed as a rule"* before the teaching experiment, mentioned that she *"likes to have a definition as brief as possible"* in the interview session.

As Hoffer stated, the ability to define words accurately and concisely is regarded as being at van Hiele level 3, "ordering", and the ability to understand the distinctions among definitions and axioms, and postulates is at level 4, "deduction" (Hoffer, 1981). The ability at these levels is certainly necessary for appreciating the important role of definitions. The analysis of the "levels of discourse" together with the changes of students' views about the nature and role of definitions suggest that in order for their concept of definition and its role in a formal system to be expanded, students need to engage in the activities across various levels of discourse.

The task used in session 8 appeared to function well for producing the need for examining the definition of exterior angle. When one is asked to define an exterior angle to an obtuse angle, there are several possibilities to consider. To make a new definition by "extending" the original one itself is one way and to hold the same definition in a broader context while keeping certain consistencies is another. In the latter case, which is often the case in mathematics and which was chosen by students S & U, the consistency between two definitions became significant when they wanted to hold the sum of exterior angles as 360°. For students S & U to maintain the consistency, the need for monitoring and verifying their work seemed to increase.

5 Final Remarks

Rich learning experiences in learning mathematics can be provided when students are involved in the process of defining, extending, and creating mathematics concepts. In this chapter, results of a teaching experiment were presented. The pairs of tenth grade students examined and created definitions of various mathematical objects, including unfamiliar quadrilaterals such as a kite and "boomerang", and raised a new definition of exterior angle to an obtuse angle when they tried to find the sums of exterior angles of these figures.

In the teaching experiment reported in this chapter students could distance themselves from the action of making a definition and through the activity of examining definitions their view on the nature of definition seemed to be changed. For providing rich learning experiences to students and helping students to appreciate the nature and role of definitions in mathematics, the constructive and tentative nature of definitions in mathematical reasoning should be emphasized through the experiences in which definitions are socially constructed with critiques from others.

References

Borasi, R. (1992). *Learning mathematics through inquiry*. Portsmouth, NH: Heinemann.

Borasi, R. (1994). Capitalizing on errors as "springboards for inquiry": A teaching experiment. *Journal for Research in Mathematics Education*, 25(2), 166-208.

Fawcett, H. P. (1938/1995). *The nature of proof: Description and evaluation of certain procedures used in a senior high school to develop an understanding of the nature of proof*. New York: Columbia University Teachers College Bureau of Publications. (Reprinted by National Council of Teachers of Mathematics, Reston, VA: The Council)

Hoffer, A. (1981). Geometry is more than proof. *Mathematics Teacher*, 74, 11-18

Ministry of Education, Culture, Sports, Science, and Technology (2008). *The course of study*. Tokyo: The author.

Nakamura, K. (1971). An introduction to the elements. In Nakamura et al. (Eds.) *The elements*. Tokyo: Kyouritsu

Popovic, G. (2012) Who is this trapezoid, anyway? *Mathematics Teaching in the Middle School*, 18(4), 196-199.

Shimizu, Y. (1997). Defining an exterior angle of certain concave quadrilaterals: The role of "supposed others" in making a mathematical definition. In J. Dossey et al. (Eds.) *The Proceedings of the Nineteenth Annual Meeting of the North American Chapter of the International Group for the Psychology of Mathematics Education*. Illinois State University. 223-229.

Usiskin, Z., & Griffin, G. (2008). *The classification of quadrilaterals: A study of definition*. Charlotte, NC: Information Age Publishing.

Vinner, S. (1991). The role of definitions in the teaching and learning of mathematics. In D. Tall (Ed.) *Advanced mathematical thinking*. Dordrecht: Kluwer

Vinner, S., & Dreyfus, T. (1989). Images and definitions for the concept of function. *Journal for Research in Mathematics Education*, 20(4), 356-366.

Wilson, P. S. (1986). The relationship between children's definitions of rectangles and their choices of examples. In G. Lappan & R. Evans (Eds.) *Proceedings of the Eighth Annual Meeting of the North American Chapter of the International Group for the Psychology of Mathematics Education*. East Lansing, MI: University of Michigan.

Zandieh, M., & Rasmussen, C. (2010). Defining as a mathematical activity: A framework for characterizing progress from informal to more formal ways of reasoning. *Journal of Mathematical Behavior*, 29, 57-75.

Zazkis, R., & Leikin, R. (2008). Exemplifying definitions: A case of a square. *Educational Studies in Mathematics*, 69, 131-148.

Chapter 7

Teaching for Abstraction through Mathematical Learning Experiences

CHENG Lu Pien

Learning experiences influence the ways teachers teach and students learn. In this chapter, examples of learning experiences within three mathematics lessons in the primary school are used as a basis for considering how teaching for mathematical abstraction may look. The students' reflections, teachers' critique of the three lessons and researchers' field notes were used to identify critical features that define worthwhile learning experiences within the three mathematics lessons. The analysis of the lessons and reflections from students and teachers suggests how learning experience may be designed to promote mathematical abstraction.

1 Introduction

There are many conceptions of mathematics identified in mathematics education literature. Dossey (1992) reported three conceptions of mathematics "as a personally constructed, or internal, set of knowledge" (p. 44). One of these three conceptions positions mathematics as a process, that is, "knowing mathematics is equated with doing mathematics" (p. 44). In this view, Dossey reported that "the learners develop their own personalized notions of mathematics as a result of the activities in which they participate" (p. 44). It is the "doing" that forms the mathematics. Here, Dossey refers "doing" to "experimenting, abstracting, generalizing, and specializing" (p. 44). This conception of mathematics is in accord with the 2012 Singapore Primary School Mathematics Syllabus where students are engaged in learning experiences and these learning experiences are descriptions of actions that students will "do" and activities that students will go through in order to achieve curricula goals. The second of the three conceptions views "mathematics as a result from social interactions". In this view, mathematics is learnt when the learner acquired the "relevant facts, concepts, principles, and skills as a result of social interactions that rely heavily on context" (Dossey, 1992, p. 45).

One of the key features of the 2012 Mathematics Primary Syllabus is the "explication of learning experiences" (p. 6). The learning experiences "describe actions that students will perform and activities that students will go through" so that curriculum objectives can be achieved (MOE, 2012, p. 20). The range of learning experiences includes (and not limited to) the use of concrete materials, hands-on activities, real-world context, problem posing, and using calculators. Of particular interest in this chapter is the use of learning experience to help students "relate mathematical concepts with concrete experiences" (Ministry of Education, 2012, p. 15) and problem posing to enhance conceptual understanding.

Utilizing the theoretical framework developed by White and Mitchelmore (2010), this chapter examines three mathematics lessons to show how teachers can utilize a range of learning experiences that support abstraction.

2 Teaching for Abstraction

Bruner (1966) suggested that the stages of learning can involve enactive, the iconic, and the symbolic experiences. According to Bruner, (cited in Orton, 2004, p. 178) the first stage of learning an abstract concept, for example place value concept, requires an enactive approach where children are engaged in manipulating concrete materials. The next stage to learning involves the use of visual images so that eventually the actual concrete objects become iconic representations of the objects. The third stage to learning is symbolic. The three stages follow a sequential approach. In the classroom situation this typically involves concrete experiences followed by pictures, followed by pencil-and-paper tasks (Orton, 2004). The concrete-pictorial-abstract (C-P-A) approach used since the early 1980s in the Singapore primary mathematic curriculum can be equated to Bruner's stages of learning. However, one concern that is also shared by White and Mitchelmore (2010) is that students often learn "the fundamental mathematical concepts in isolation, without engaging in the abstraction process" (p. 207). In other words, how do we derive maximum benefit from the use of concrete experiences so that the intended mathematics is taught effectively through the use of a particular learning experience? With reference to concrete materials, Sharma (1987) suggested that explicit links "between meaningful actions on the objects and its translation into related symbols" (p. 134) is needed.

According to Skemp (1986)

> *Abstracting* is an activity by which we become aware of similarities among our experiences. *Classifying* means collecting together our experiences on the basis of these similarities ... the result of abstracting, which enables us to recognise new experiences as having the similarities of an already formed class. To distinguish between abstracting as an activity and abstraction as its end-product, we shall call the latter a *concept*. (p. 21)

Drawing from Skemp's work, the first step for mathematical abstraction to take place is the opportunity to engage in learning experiences. Next, similarities among the experiences need to be highlighted before collecting these experiences on the basis of these similarities (classification) so that when pupils are faced with new experiences and situations, they are able to recognize the new experiences as having the similarities of an already formed class. When the pupil achieves this (end-product/abstraction) the student has acquired the concept.

The teaching for abstraction framework developed by White and Mitchelmore (2010) draws from Skemp's (1986) empirical abstraction to propose a four-phase model of teaching for abstraction. The four teaching phases that White and Mitchelmore (2010, p. 209) include:

- Familiarity: Provide students a variety of contexts for students to explore. The concept arises through the exploration of the contexts (form generalizations about individual contexts and become familiar with the underlying structure of each context).
- Similarity: Help students recognise the similarities and differences between the underlying structures of these contexts.
- Reification: Help students draw out the general principles underpinning the similarities in the context. Support students to abstract the intended concept.
- Application: Give new situations for students to use the concept.

In this chapter, I use the framework as a lens to re-examine mathematics lessons that teach for abstraction in the primary school.

3 This Study

This is a qualitative case study conducted in two research sites involving two different groups of teachers that employed laboratory class cycle as their professional development tools. In each cycle of the laboratory class, the researcher worked with a group of teachers (about 7 teachers) in the same school to plan, teach, observe and critique a mathematics lesson. The teachers identified the mathematics topics to be studied for the laboratory class. The laboratory class cycle has the same structure of investigation (planning, research lesson and reflection) as the lesson study cycle (Lewis, Perry, & Hurd, 2009) except that in the laboratory class cycle, the lessons are not repeated. Three research lessons resulting from the laboratory class cycles were selected for analysis.

3.1 *Purpose of study*

The study seeks to provide examples of learning experiences that enhanced mathematical abstraction. The research questions that guided this study are:

(1) What does teaching for abstraction looks like within the lessons?

(2) What makes the learning experiences in the three mathematics lessons worthwhile?

3.2 *Selection of lesson*

According to Cockcroft (1982) and Stigler and Hiebert (1999), mathematical learning involves five distinct purposes mainly, developing fluency when recalling facts and performing skills, interpreting concepts and representations, developing strategies for problem solving, awareness of the nature and values of the educational system, and an appreciation of the power of mathematics in society. This is aligned to Singapore's pentagonal framework of the mathematics curriculum. In this chapter, we focus primarily on the learning of concepts - one of the five components underpinning the teaching and learning of mathematics in the primary school. In selecting the lessons for analysis, this emphasis on concept formation and enhancement was to the fore. Other considerations included evidence of opportunities for students to be involved in at least two of the following experiences: hands on activities, use of manipulative or concrete objects, or creating word problems.

3.3 *Data analysis*

To answer the first research question, each of the three mathematics lessons were analyzed using the teaching for abstraction framework. Data available for analysis included a photographic record of key teachers' actions, audio recordings of the meetings, teaching materials, samples of students' work and researchers' field notes. Key words exemplifying each stage from the literature were used (e.g. variety of context, structures in each context) in order to identify the stages in each of the lesson. The codes were expanded from the data that contribute to the attainment of the stages (e.g. familiarity of prior knowledge, familiarity across contexts, familiarity within one context).

To answer the second research question, students' reflections in their worksheet, teachers' critique and feedback about the lessons were compared. From the comparison, common themes were identified that characterises the worthwhileness of the learning experiences in each of the three mathematics lessons.

In the following paragraphs, descriptions of how mathematical abstraction can be achieved in the lessons are provided. The framework was extended and modified using the codes to cater to the needs of the students.

4 The Mathematics Lessons

4.1 *Measurement lesson*

This lesson focused on the metric conversion from metre to centimetre and took place in research site 1. A total of 7 teachers teaching the upper primary mathematics participated in the professional development program involving six sessions. Four sessions were used to plan the lesson, one lesson was used to teach and observe the lesson and the last session was used to critique the lesson and make further changes to the lesson plan. The lesson was taught to an intact class of 36 Primary 5 students. The activity involved students working in pairs to estimate and measure 6 objects in the classroom using both a measuring tape and metre ruler. A worksheet was designed for the students to record their estimation and actual length of the objects in centimetres. They were then required to convert the measurement from centimetres to metre.

Familiarity

In this phase, the students familiarized themselves with the metre ruler and measuring tape to measure the objects assigned to them. The objects to be measured were categorized into 3 boxes to help pupils develop a sense of length more than 1 metre and length less than 1 metre. During this phase, students' errors in conversion surfaced in the students' work and these errors were used as teaching points in teaching Phase 2 and 3. Samples of students' errors are shown in Figures 1 and 2.

Box 1			
Item	Predictions / Guess (in cm)	Actual Measurement (in cm)	Convert each measurement to metres
Height of steel table	51 cm	74 cm	740
Height of chair	63.5 cm	~~74 cm~~	

Box 2			
Item	Predictions / Guess (in cm)	Actual Measurement (in cm)	Convert each measurement to metres
Height of TV rack from the floor	145 cm	122 cm	1.22 m
Length of teacher's table	95		
Length of 5 square tiles	125 cm	59 5	59.5 m

Figure 1. Sample of student's work (wrong conversion)

Box 2			
Item	Predictions / Guess (in cm)	Actual Measurement (in cm)	Convert each measurement to metres
Height of TV rack from the floor	1.45	122	1.22 m
Length of teacher's table	98		
Length of 5 square tiles	125	59.5	59.5 m

Figure 2. Sample of student's work (wrong estimation and conversion)

Similarity

In this phase, the teacher helped the pupils to recognize the similarities and differences between the objects in the 3 boxes. In the first box, the objects are each less than 100 cm, hence, the readings falls within one metre ruler when measuring the objects in Box 1. In the second box, the objects measure between 100cm and 200cm, hence between 1 and 2 metre rulers were needed to measure objects in the second box. In the third box, the objects measure between 200cm and 300cm, hence between 2 and 3 metre rulers were needed to measure objects in the third box.

Reification

In this phase, the teacher helped the pupils to draw out the mathematics from Phase 2. In box 1, e.g. height of the steel table is 74cm, measures less than 1 metre ruler, hence the steel table is 74 out of 100 cm. The students' pre-requisite knowledge was used to convert a fraction to decimal, 74/100=0.74m. In box 2, e.g. height of TV rack is 122 cm. This requires 1 metre ruler and another 22 cm to measure the TV rack which translates to 1m and 22 cm in mathematical symbol. This measures to 1.22m. This concept is repeated using similar structures to readings in box 3.

Application

In this phase, the teacher helps the pupils to apply the concept to another situation. An example is shown in Figure 3.

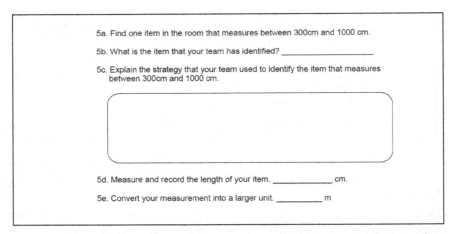

Figure 3. Application in conversion of units from smaller to bigger unit

4.2 Building multiplication table of 4 lesson

A total of 7 teachers from research site 2 planned this lesson for primary 2 children. The multiplication table of 4 was identified by the teachers as the most difficult multiplication facts to learn in primary 2. The teachers wanted to investigate effective ways of helping children to make sense of the multiplication table of 4 as well as to commit it to memory. The group used 3 sessions to plan the lesson, 1 session for observing and 1 session for critique. The lesson aimed to help pupils build the multiplication table of 4 and use the skip-count in fours strategy to find the four facts. Students explored the interlocking blocks to build groups of 4 and were exposed to the concrete-pictorial-abstract approach for this lesson. The Research Lesson reported here is the revised lesson based on the teachers' critique of the public lesson.

Familiarity
In this phase, the teachers identified some pre-requisite knowledge and understanding as playing a major role in building the multiplication table of 4. The pre-requisites included concepts of equal groupings and skip-counting in 4 strategy. Also, the lesson deliberately used familiar contexts involving the concept of equal groupings (e.g. a box with 4 sweets, a car with 4 wheels). Figure 4 shows students working in groups and taking turns in their groups to form the multiplication table of 4 with the interlocking cubes.

Figure 4. Concrete (building of 4 times table)

Mathematical terms 'groups of 4' were repeated to help students become familiar with the underlying structure of the multiplication table of 4.

Similarity
In this phase, the teacher showed the pupils a series of pictures that develops the multiplication table of 4. In building the multiplication table of 4 using the pictures, similarities and differences between the pictures

were discussed. The students were asked to observe how the pictures or situations differ and how are they similar. Mathematical terms 'groups of 4' were repeated in every picture to help students become familiar with the language to describe the underlying structure of multiplication table of 4. Figure 5 shows the repeated use of key mathematical terms to describe the multiplication table of 4.

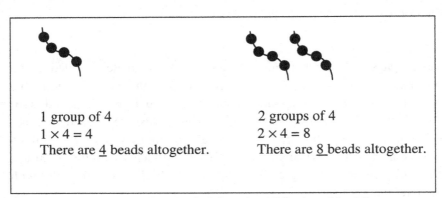

1 group of 4
$1 \times 4 = 4$
There are <u>4</u> beads altogether.

2 groups of 4
$2 \times 4 = 8$
There are <u>8</u> beads altogether.

Figure 5. Pictorial (building of multiplication table of 4)

Reification
In this phase, the teacher helped the pupils to draw out the structures underlying the multiplication table of 4, that is, concept of equal grouping, adding 4 each time, linking and connecting the skip count in 4 strategy to find the four facts.

Application
In this phase, the teacher may help the pupils to apply the concept to a word problem. For example, Tammy has 5 boxes. Each box has 4 sweets. How many sweets does Tammy have altogether?

4.3 *Problem posing lesson*

The same group of teachers who conducted the measurement lesson was involved in the lesson for problem posing. The same amount of time was spent on each stage of the laboratory class cycle for this lesson as the measurement lesson. The lesson involved the teacher revising the concept of fraction division using the fraction circles to model a story sum for ½ ÷ 1/8. The students next worked in pairs to pose a story sum for 2/3 ÷ 1/6. They then chose their own proper fractions to create a story sum for fraction divided by fraction. Different from the rest of the laboratory class cycles, one of the teachers in the group (Teacher 1) conducted the Research Lesson separately on her own with her students before observing the scheduled Research Lesson with the rest of the group members. The rest of this section describes an example of how problem posing may support teaching for mathematical abstraction in the mathematics classroom based on the feedback provided by Teacher 1 and the teachers' critique of the public lesson.

Familiarity
In this phase, the students were already exposed to various contexts where fraction division can be used. They were familiar with the structure 'how many [divisor] in [dividend]'. They made story sums, solved their own story sums and gave their story sums to another group of students to solve. Figure 6 shows samples of word problems posed by students. During this phase, students' errors in posing a correct word problem surfaced and these errors can be used in teaching Phase 2 and 3.

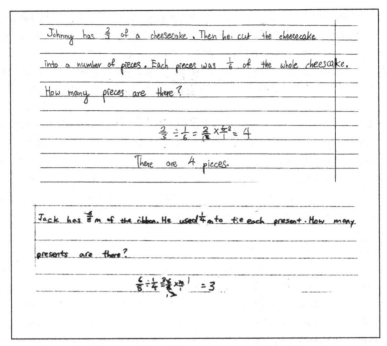

Figure 6. Sample of students' problems

Similarity

In this phase, the teacher helped the pupils to recognize the similarities and differences between the word problems posed by the various groups. In one of the categories of responses, the students posed word problems using measurement concepts. Fewer errors were found when the measurement context was used to pose fraction division problems. Another category of responses showed students posing word problem using discrete items (e.g., a cake, a pizza). This category of responses gave rise to many problems. One of the problems was associated with the use of language, for example, in the 2-step word problem in Figure 7, the word 'remaining' caused some confusion for the students. Another problem concerns the referent whole, for example, whether the *pizza* in '1/8 of *pizza*' referred to 1/8 of the whole pizza or 1/8 of ¾ of the pizza for the 2-step word problem.

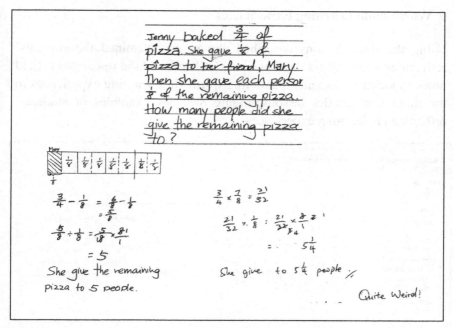

Figure 7. Sample of students' problems and solutions

Reification

In this phase, the teacher helped the pupils to add greater clarity to the generalizations on the concept of fraction division especially when in different context (measurement and discrete). Understanding the structure 'how many [divisor] in [dividend]' is insufficient. Deeper understanding of the concept required students to be able to identify the divisor and the dividend. The concept of 'where is the whole' was again identified as the key factor in understanding fraction division concept and the importance of 'accurate' language used to describe fraction division situations was also affirmed.

Application

In this phase, the teacher helped the pupils to apply the division concept to either create more word problems for fraction division or correct/reword the word problem to fit the intended solution.

5 Worthwhile Learning Experiences

Using the three lessons described above, we examined the students' reflections, the teachers' critique and feedback, and the researchers' field notes to identify elements that define worthwhile learning experiences in the three mathematics lessons. Figure 8 shows examples of students' reflections in learning experiences.

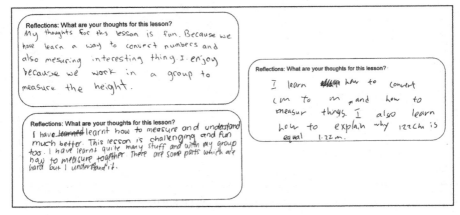

Figure 8. Students' reflections

Students found the learning experiences worthwhile when they learnt a specific concept (e.g., learnt how to measure, able to explain the mathematics) and were able to complete the mathematical task assigned to them. Opportunities for group work added to the enjoyment of the experience. For both the upper and lower primary, the teachers found the learning experiences worthwhile when their pupils engaged in and enjoyed the lesson and were able to complete the worksheets and apply the acquired concept to problem solving.

The *student-centred activities* offered opportunities for students to be pro-active in their own learning as they explored the cubes, constructed their own word problems and used the metre ruler to measure objects in the room. Teacher 2 said that for the measurement research lesson, "They [the pupils] are more engaged because it is not just frontal teaching. It is activity centred because everyone contributed. I think they are more interested". Teacher 3 said that the problem posing research lesson "offers children opportunity to make sense of the mathematical operation, numbers, the answer and context e.g. how to have ½ a person?". Teacher 3 also felt that the problem posing activity required careful selection of fractions, changing the fractions or even the context for the solution to make sense and that the whole activity offered opportunities for students to self-regulate their mathematical thinking. Learning of mathematics becomes personalized; students developed personal notions of the mathematics as a result of these activities.

Opportunities for *group work* and *opportunities to communicate* (through singing, explaining, problem posing) their mathematical ideas contributed to the students' enjoyment of the learning experiences. Across the three lessons, the effective use of *scaffolding questions* was identified to be key in helping pupils achieve mathematical abstraction. Scaffolding questions that helped pupils identify similarities and differences between underlying structures of the context helped pupils see patterns, relationships and structures. Teacher 4 shared that "with the amount of guidance [scaffolding] given during the lesson, there is no problem with the students completing the worksheet". Teacher 4 also added that the effort to plan meaningful learning experiences makes it all worthwhile when the pupils' enjoyment enhanced their own enjoyment of the lesson. Reflecting on the lesson on building multiplication table of 4, Teacher 4 added, "The topic is fun because there are a lot of songs, movements and activities in this lesson ... the class environment is such that the students know it is alright to make mistakes and we learn from our mistakes".

6 Further Implications for Teaching

Based on the above findings, the teaching for abstraction framework can be a useful and relevant framework for planning and designing lessons for concept formation leading to problem solving. We applied and extended the framework in the familiarity phase to include familiarity with the pre-requisite knowledge and any necessary tools or materials. In some instances, concrete materials can be used to represent a variety of context and concepts are developed through the use of manipulatives.

The framework was usefully applied to problem posing lessons. We see problem posing as an activity that helps pupils further generalize mathematics concepts by embracing more contexts.

The framework provides teachers a lens to re-examine and plan mathematics lessons that teach for abstraction. The framework is also useful in the planning of our mathematics lessons for concept formation and problem solving. For example, one of the learning difficulties for upper primary students is identifying the correct base in percentage. Mathematical abstraction of this concept may be achieved through a series of learning experiences (see Appendix) that embed the 4 stages of teaching for abstraction and what we identified to be worthwhile learning experiences. They were as follows:

- Familiarity – pupils becomes familiar with the change in total as they add and remove the beads. Each time when beads are removed or added the percentage for each of the colored beads is calculated.
- Similarity – pupils are challenged to think about the differences and similarities through the scaffolding question in the activity worksheet. E.g. in each instance when the beads are removed, what happened? What remained the same and what has changed? Where is the whole and where is the part? What happens to the percentage of beads for each of the coloured beads when beads are removed?
- Reification – draw similarities to part-part whole relationships. When a part changes, what happens to the whole? How do you calculate this change in terms of percentage?
- Application – apply this concept to a word problem.

The examples in this chapter are intended to illustrate how teaching for abstraction may look like in the primary school mathematics class. These examples can assist teachers to think about what teaching for abstraction looks like as they design learning experiences to achieve curriculum objectives.

References

Bruner, J. S. (1966). *Toward a theory of instruction*. Cambridge, MA: Harvard University Press.

Cockcroft, W. H. (1982). *Mathematics counts*. London: HMSO

Dossey, J. (1992). The nature of mathematics: It's role and its influence. In D. A. Grouws (Ed.), *Handbook of research on mathematics teaching and learning* (pp. 39-48). New York: Macmillan.

Lewis, C., Perry, R., & Hurd, J. (2009). Improving mathematics instruction through lesson study: A theoretical model and North American case. *Journal of Mathematics Teacher Education, 12*, 285-304.

Long, L. (2003). *Delightful decimals and perfect percents: Games and activities that make math easy and fun*. NJ: John Wiley & Sons.

Ministry of Education (Singapore) (2012). Primary mathematics teaching and learning syllabus. Downloaded
http://www.moe.gov.sg/education/syllabuses/sciences/files/maths-primary-2013.pdf

Orton, A. (2004). *Learning mathematics: Issues, theory and classroom practice* (3rd ed.). London: Continuum.

Sharma, M. (1987). How to take a child from concrete to abstract. *Math notebook*. ERIC Document reproduction Service No. ED 342600.

Skemp, R. R. (1986). *The psychology of learning mathematics* (2nd ed.). Harmondsworth, UK: Penguin.

Stigler, J. W., & Hiebert, J. (1999). *The teaching gap: Best ideas from the world's teachers for improving education in the classroom*. New York: Free Press.

White, P. & Mitchelmore, M. C. (2010). Teaching for abstraction: A model. *Mathematical Thinking and Learning, 12*, 205-226.

Appendix

Beads – Colours & Number*

Activity 1

1. Open the bag of beads.
2. Sort the beads by colours and complete the table below.

Colours	Number	Percent
Red		
Orange		
Yellow		
Brown		
Green		
Total		

(a) How many beads are there?   _____ beads
(b) What percentage of the beads is red?   _____ %
(c) What percentage of the beads is orange?   _____ %
(d) What percentage of the beads is yellow?   _____ %
(e) What percentage of the beads is brown?   _____ %
(f) What percentage of the beads is green?   _____ %

3. You are given 5 **more red** beads.

   Complete the table below.

Colours	Number	Percent
Red		
Orange		
Yellow		
Brown		
Green		
Total		

4. Compare the two tables.
    Write down your observations in the space below.
5. Word Problem
    In a bag of 32 beads, 25% of the beads were red. The rest of the
    beads were green. You are given some more red beads and 40%
    of the beads were red. How many red beads were you given?

Activity 2

1. Use all the beads from *Activity 1*.
2. Sort the beads by colours and complete the table below.

Colours	Number	Percent
Red		
Orange		
Yellow		
Brown		
Green		
Total		

(a) How many beads are there?   _____ beads
(b) What percentage of the beads is red?   _____ %
(c) What percentage of the beads is orange?   _____ %
(d) What percentage of the beads is yellow?   _____ %
(e) What percentage of the beads is brown?   _____ %
(f) What percentage of the beads is green?   _____ %

3. Remove **5 brown** beads.
    Complete the table below.

Colours	Number	Percent
Red		
Orange		
Yellow		
Brown		
Green		
Total		

4. Compare the two tables.
    Write down your observations in the space below.

5. Make a necklace using all the beads.
6. Word Problem
    In a bag of 32 beads, 25% of the beads were red. The rest of the beads were green. Some red beads were removed and 20% of the beads were red. How many red beads were removed?

Activity 3

1. Open the bag of beads.
2. Sort the beads by colours and complete the table below.

Colours	Number
Red	
Green	
Total	

(a) How many beads are there? _____ beads
(b) What percentage of the beads is red? _____ %
(c) What percentage of the beads is green? _____ %

3. **Remove** 10 **red** beads.
    (a) How many red beads did you remove? _____ beads
    (b) What is the percentage decrease in the **red** beads? _____ %
    (c) What is the percentage decrease in the beads? _____ %
    (d) How many red beads are there now? ___ red beads
    (e) How many green beads are there now? ___ green beads
    (f) How many beads are there now? ___ beads
    (g) What percentage of the beads is green? ___ %
    (h) What percentage of the beads is **red**? ___ %
    (i) What conclusions can you draw about the beads?

4. You are given 15 **more green** beads.
    (a) How many more green beads were you given? _____ beads
    (b) What is the percentage increase in the **green** beads? _____ %
    (c) What is the percentage increase in the beads? _____ %
    (d) How many red beads are there now? _____ red beads
    (e) How many green beads are there now? _____ green beads
    (f) How many beads are there now? _____ beads
    (g) What percentage of the beads is **green**? _____%

(h) What percentage of the beads is red?   ____%

(i) What conclusions can you draw about the beads?

5. Make a necklace using **all** the beads.
   (a) How many beads are there in the bag now?   _____ beads
   (b) What percentage of the necklace is red?   _____ %
   (c) What percentage of the necklace is green?   _____ %

6. Word Problem

   There were 32 red and green beads in a bag. 2 red beads were removed. The number of green beads was increased by 25%. The bag now has 36 beads. How many red beads were in the bag at first?

* The above series of activities are adapted from Long (2003).

Chapter 8

Making Sense of Number Sense: Creating Learning Experiences for Primary Pupils to Develop their Number Sense

YEO Kai Kow Joseph

In the primary levels, learning what numbers mean, how they may be represented, relationships among them and computations with them are central to developing number sense. Number sense enables primary school pupils to understand and express quantities in their world. Researchers indicate that number sense develops gradually, and varies as a result of exploring numbers, visualizing them in a variety of contexts, and relating them in ways that are not limited by traditional algorithms. This chapter describes the concept of number sense, the importance of number sense, a framework for examining basic number sense and the various learning experiences to promote number sense. Six Learning Experiences are highlighted that teachers might trial in their mathematics lessons to engage pupils in making sense of numbers.

1 Introduction

A primary 2 pupil in a mathematics lesson was being observed. After writing the addition equation 69 + 16 in vertical form and drawing a horizontal line, the primary 2 pupil wrote the answer 85.

Observer: Wow. Well done. Tell me how you did that.

Pupil: OK, (answered the pupil reluctantly) but don't tell my teacher. I know how to add 69 and 10. I will get 79 and then I count on 6 in my head to get 85.

Observer: That is a very good way. Why can't I tell your teacher?

Pupil: Because I must show working but I can't understand
how to carry over, so I do my way in my head and get
the answer.

This primary 2 pupil could not follow the formal written algorithm but understood enough about whole number to invent her own efficient mental strategy. One might say that the primary 2 pupil has strong number sense in relation to addition of whole number. The progress of number sense is of international concern and interest. Mathematics educators are concerned that many pupils show little understanding of numerical situations when they have to solve whole number problems. Although the significance of understanding may well have "long been recognized" in theory, the term number sense is a relatively new idea in the primary school mathematics curriculum. Several number sense related international research studies have shown that pupils at primary levels are lacking of number sense (Alajmi, 2004; Markovits & Sowder, 1994; Menon, 2004; Yang, 2005; Yang, Hsu, & Huang, 2004). Mathematics teachers need to be wary that number sense is not a discrete set of skills to be taught over a short period of time or a set of skills for high-ability pupils. It must be part of pupils' daily mathematical lessons that slowly grows and develops over time. This chapter therefore defines the concept of number sense, discusses the importance of number sense, examines a framework involving number sense, and shares learning experiences to promote number sense at the primary level.

2 Review of Literature

2.1 *Number sense*

What is number sense? There are differing perspectives to consider when searching for a sufficient and workable definition of number sense. Moreover, no two mathematics educators in the research field define the concept of number sense in a similar way (Berch, 2005). Although most mathematics teachers and educators can effortlessly identify number sense when they encounter it, defining what it is and how it can be taught is challenging. In the National Statement on Mathematics for Australian Schools number sense is defined as "ease and familiarity with and intuition about numbers" (Australian Education Council, 1991, p. 107). In the same vein, Howden (1989) agreed that pupils who have number sense would have good intuition about numbers and their relationships. It develops gradually as a result of discovering numbers, representing them in a variety of contexts, and relating them in ways that are not limited by traditional procedures.

Number sense is often considered as "flexibility" and "inventiveness" (Dunphy, 2007, p. 2). Sowder (1992) indicated that number sense "represents a certain way of thinking rather than a body of knowledge that can be transmitted to others" (p. 3). Furthermore, Sowder (1992) listed eight behaviors that are indicative (but not proof) of the existence of number sense. These relate to flexible understandings of number representations and the ability to provide appropriate representations, effective use understandings of both the relative and absolute magnitudes of numbers, meaningful connections between operations and symbols, use of number benchmarks and flexible strategies for mental computation and estimation, and an inclination to make sense of numbers. According to Schneider and Thompson (2000), number sense includes a solid understanding of the meaning of numbers and numerical relationship; flexible thinking about number (for example, being able to see one hundred in a variety of ways). Later research studies on number sense have used number sense to refer to a person's general understanding of numbers and operations and his/her ability to handle daily-life situations that include numbers. This ability demands the usage of useful, flexible, and efficient strategies, such as mental computation and estimation, to grasp numerical problems (McIntosh, Reys & Reys, 1992; Reys & Yang 1998 ; Sowder, 1992; Yang, Hsu, & Huang, 2004).

2.2 *Importance of number sense*

The development of number sense is important in mathematics education. Because of this, number sense has generated many discussions and studies by mathematics educators, researchers and educational psychologists (Greeno, 1991; Markovits & Sowder, 1994; McIntosh, Reys, & Reys, 1992; Siegler & Booth, 2005; Verschaffel, Greer, & De Corte, 2007). Moreover, many studies have shown the importance of number sense for primary school pupils (Doig, McCrae, & Rowe, 2003; Gersten & Chard, 1999; Griffin, 2004; Griffin, Case, & Siegler, 1994). Before children go to school, they begin to exhibit sophisticated informal number sense that often goes unnoticed by teachers. Children who have developed early number sense are able to recognize small amounts, demonstrate basic counting skills and add small quantities. In addition, early number sense is mostly intuitive, developing from a range of daily play experiences. Pupils with limited number sense are considered as not progressing beyond early counting skills or strategies, relying on procedural knowledge to meet the demands of numeracy in their environment, and having difficulty managing the demands of mathematical language (Evans, Strnadová, & Wong, 2007).

Why is the teaching and learning of number sense for primary school pupils so important? Firstly, many mathematics teachers and educators would agree that number sense, or the capacity to make sense of numbers and quantity, is foundational knowledge required by pupils. It assists pupils to link quantities in the number system and to develop mathematical strategies. Primary school pupils with number sense develop "meaning" for numbers and their relationships. In fact, the National Council of Teachers of Mathematics, in their Principles and Standards for School Mathematics, state that number sense is one of the foundational ideas in mathematics. They state that pupils with number sense (1) Understand number, ways of representing numbers, relationships among numbers, and number system; (2) Understand meanings of operations and how they related to one another; (3) Compute fluently and make reasonable estimates (NCTM, 2000, p. 32). Secondly, overemphasis on written computation often impedes the pupils' mathematical thinking and understanding (Burns, 1994; Kilpatrick, Swafford, & Findell, 2001; Reys & Yang, 1998). Many pupils are good rule followers but may not have relational understanding of the techniques which they have learned (Hiebert, 1986; McIntosh, et al., 1997). According to Gersten and Chard (1999), the construct of number sense is critical to conceptual understanding. Woodard and Baxter (1997) explain that if number sense is not evident in pupils, they will face many challenges in many arithmetic topics, especially those related to conceptual understanding and application of mathematical concepts to new problem-solving scenarios. Thirdly, number sense should "be a holistic concept related to everyday use of number and to encompass skills, understanding, disposition, and flexibility" (Dunphy, 2007, p. 8). It has been widely discussed and generally accepted that developing number sense involves meaningful learning and understanding (Anghileri, 2000; Kilpatrick, Swafford, & Findell, 2001; NCTM, 2000; Yang, 2005). Research has shown that through the provision of rich learning environments, where explorations of number combinations and arrangements are encouraged, pupils spontaneously develop their own strategies for basic fact combinations (Fuson, 1992) as well as improve number sense through exploration of number relationships (Wright, 1996). Teaching and learning number sense should therefore be emphasized in primary school mathematics lessons. Number sense needs to be recognized, understood, accommodated and taught effectively.

2.3 A framework for number sense

McIntosh, Reys and Reys (1992) believed that many teachers have an inaccurate opinion of their pupils' level of number sense and most teachers do not recognize this mismatch between their beliefs and practices. More than two decade ago, McIntosh, Reys and Reys (1992) developed a framework for considering the development of number sense. It was based on the literature associated with number sense and comprised three broad categories:

• Knowledge of and facility with numbers,
• Knowledge of facility with operations,
• Applying knowledge of and facility with numbers and operations to computational settings.

An overview of the framework is presented in Table 1 (McIntosh, Reys, & Reys, 1992). The framework articulates a scheme that elucidates, organizes and interconnects some of the generally agreed upon categories of number sense.

Table 1

Overview of framework for considering number sense (McIntosh, Reys & Reys, 1992)

Broad Categories	Understandings
Knowledge of and facility with numbers	• Sense of orderliness of numbers • Multiple representations for numbers • Sense of relative and absolute magnitude of numbers • System of benchmarks
Knowledge of facility with operations	• Understanding the effect of operations • Understanding mathematical properties • Understanding the relationship between operations
Applying knowledge of and facility with numbers and operations to computational settings	• Understanding the relationship between problem context and the necessary computation • Awareness that multiple strategies exist • Inclination to utilize an efficient representation and/or method • Inclination to review data and result for sensibility

McIntosh, Reys and Reys (1992) also identified six strands (two from each of the three major categories of the framework). The six strands are:
1. Number Concepts
2. Multiple Representations
3. Effect of Operations
4. Equivalent Expressions
5. Computing and Counting strategies
6. Measurement of Benchmarks

3 Learning Experiences and Number Sense Strands

With the introduction of 2013 Mathematics syllabus in Singapore, it is appropriate to revisit the six number sense strands (McIntosh, Reys, & Reys, 1992) in terms of their relevance and applicability to the teaching of primary mathematics. In the Singapore Revised Mathematics Syllabus (Ministry of Education, 2012), one of the main emphases of the primary level mathematics curriculum is the explication of learning experiences. While the revised curriculum continues to emphasize learning experiences in the classroom, there is now an even greater focus on the development of pupils' number sense. Anecdotal evidence from many experienced mathematics teachers suggests the development of number sense is best if the focus is consistent and happens regularly during mathematics lessons. Number sense is not a finite entity that a pupil either has or does not have but rather a process that develops and matures with experience and knowledge (McIntosh, Reys, & Reys, 1992). Reys (1994) pointed out that teaching for the development of number sense requires conscious, coordinated effort on the part of the teacher to help students to build connections and meaning. Moreover, mathematics teachers need to have a well-developed and meaningful understanding of the number system themselves and also pedagogical practices that provide learning experiences for pupils to explore and construct ideas about numbers. Since appropriate learning experiences in the classroom are a key factor, it is helpful that the framework and the six number sense strands (McIntosh, Reys & Reys, 1992) provide a structure for designing learning experiences in the classroom. The following section describes six learning experiences that can be incorporated in the teaching and learning of mathematics at the primary level. The six learning experiences will assist the mathematics teacher to focus less on algorithms and more directly on developing number sense.

3.1 *Number sense strand 1 – Number concepts*

Learning Experience 1, shown in Figure 1, is the place value game. It is a valuable learning experience for primary school pupils who are still trying to make sense of the structure of the number system. A good understanding of the number system assists the pupil to compare and order numbers faced with a mathematical situation. The game involves pupils using a place value board and number cards to develop an understanding of the structure of the number system and to learn to operate on numbers using this structure. By encouraging the overall winners to share their winning strategies, pupils can realize that other than chance, knowledge of place value is an aid to winning a game. This learning experience particularly develops pupils' confidence with number magnitude, order and operations.

Learning Experience 1: Place Value Game (Primary 3)
Who has a bigger difference?
Number of players: two
Materials: A set of number cards numbered 0 to 9 and a place value
        board for each pupil.
How to play:
1. Each pupil will shuffle their number cards.
2. Players stack their cards in a pile face down in front of them.
3. The two players will then take turns to draw one card at a time from the stack and form two 4-digit numbers.
4. As the cards are drawn, the player decides which 'place' the number card should be placed on the place value board.

Place Value Board			
Thousands	Hundreds	Tens	Ones

	Thousands	Hundreds	Tens	Ones
First Number				
Second Number				
Difference of two numbers				

5. As the cards are drawn, the player decides which 'place' the number card should be placed on the game board.
6. Once the 'place' has been decided, the card cannot be removed.
7. After each player has formed the two 4-digit numbers with the cards, he/she will then find the difference of the two numbers.
8. The game ends when the players have calculated the difference.
9. The player with the bigger difference wins.
10. The overall winner is determined by playing three games.

Figure 1. Learning experience 1: Place value game

3.2 *Number sense strand 2 – Multiple representations*

Numbers appear in different situations and have multiple representations. Pupils who have number sense recognize that numbers could be represented in different forms and manipulated in many ways to explain a particular situation. Since a particular mode of representation cannot embody a number completely, it is necessary to have more than one representation. Buxton (1997) mentioned three modes: language, spatial, and symbolic.

In Learning Experience 2, shown in Figure 2, four commonly used modes of representation, namely symbol, word, diagram, and number line are reinforced. The activity is an attempt to translate these modes of representation into a logical and useful technique for teaching numbers. The technique emphasizes relationships among the different modes of representation, thus deepening understanding about number. It focuses on a variety of experiences to make the learning of numbers more meaningful and stimulating to the pupils. For example, if the decimal, 0.18, is chosen, pupils represent their understanding of 0.18 by completing each part in the representation board: write in decimal fraction ($\frac{18}{100}$), write in words (18 hundredths), shade 18 small squares and mark a cross between 0.1 and 0.2 on the number line. The quality of understanding about number is determined by the linkages among the different modes of representations of the same numerical value. The inability to fluently link the different representations together is a sign of weak understanding. In addition, teachers should experiment with representation boards using different numerical values to get a feel of what works "best" for their pupils.

3.3 *Number sense strand 3 – Effect of operation*

Sometimes teachers unduly pressurize pupils to remember the rules of placing decimal points in multiplication and division computation sums. However, pupils may forget or become confused as they have no understanding of why these rules work. In such cases it is not meaningful for them to commit the rules to memory. Pupils should be strongly encouraged to use their understanding of the quantities and of the operations to think through the placement of the decimal point. The aim of Learning Experience 3 therefore is to enhance the pupils' estimation skills as a means of placing decimal points in products and quotients. Pupils should be encouraged to interpret the first problem shown in Figure 3 as "This is about 1 times of 40, so the answer is about 40." The last division problem demands an estimate. It requires 14.7236 to be thought of as "about 147 tenths" and then this is to be shared among about 70 children, so each child will receive at least two tenths.

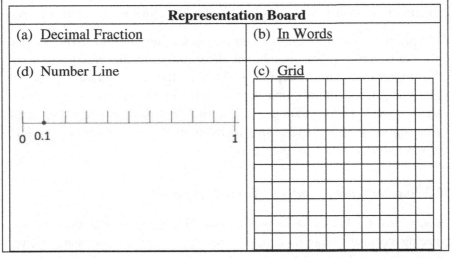

Learning Experience 2: Representing Decimals (Primary 4)
Write a decimal number less than 1 whole.
Represent the decimal number on the Representation Board
(a) Write the decimal as a decimal fraction.
(b) Rewrite the decimal number in words.
(c) Shade the squares accordingly on the grid to represent the decimal number. The large square represent one whole.
(d) Mark with a cross to show where the decimal number is on the number line.

Representation Board	
(a) Decimal Fraction	(b) In Words
(d) Number Line	(c) Grid

Figure 2. Learning experience 2: Representing decimals

Learning Experience 3: Decimal points missing (Primary 5)
Jane forgot to place the decimal point in the answer of each calculation. Describe how you can use estimation to place the decimal point correctly.

(a) 0.97 x 42 = 4 074
(b) 35.4 x 17 = 6 018
(c) 7651 x 0.0083 = 635 033
(d) 37.986 ÷ 0.004 = 44 965
(e) 14.7236 ÷ 68.3512 = 215410995

Figure 3. Learning experience 3: Decimal points missing

Pupils should also be taught to use estimation and approximation to check answers. Thus, Learning Experience 3 provides opportunities for pupils to recognise the magnitude of those numbers and the "effects of operating" on numbers, by developing "referents" for quantities and measures (Sowder & Klein, 1993, p. 41). Reflecting on the relations between the operations and numbers promotes higher order thinking and further enhances number sense (McIntosh, Reys & Reys, 1992).

3.4 *Number sense strand 4 – Equivalent expressions*

Learning Experience 4, Equivalent Expressions Snap, shown in Figure 4, illustrates how mathematical properties such as commutativity, associativity and distributivity can be used to help pupils intuitively recognize and use other equivalent expressions in a fun way. The game can be extended to include more than one operation. Pupils are normally competitive and there is a strong incentive for players to check one another's mathematical properties and challenge equivalent expressions that they think are not valid. Pupils should be given opportunities to discuss how they deduce the pair of equivalent expressions and the thinking they used. For instance the teacher could discuss with pupils how the equivalence of a pair of expressions such as 48 x 23 + 10 x 23 and 59 x 23 – 23 can be deduced. Some pupils may verbalize one of the expressions as 48 groups of 23 plus 10 groups of 23 to be the same as 58 groups of 23 while the other expression as 59 groups of 23 minus one group of 23 to be the same as 58 groups of 23.

Learning Experience 4: Equivalent Expressions (Primary 6)
Game – Equivalent Expression Snap
Number of players: two
Materials: 20 pairs of equivalent arithmetic expression cards
How to play:

1. Write a pair of equivalent arithmetic expressions on two cards.
2. You should have 20 pairs of equivalent expressions.
   For example:

   | 10 x 12 | Pair with | 9 x 12 + 12 |

   | 48 x 23 + 10 x 23 | Pair with | 59 x 23 - 23 |

1. The cards are shuffled and all are laid down with the expressions facing down in a stack.
2. Each player takes turn to open up two cards.
3. If the expressions on the two cards are equal, the player gets to keep the two cards.
4. He/she can continue to open another two cards.
5. The turn is passed onto the next player if the expressions on the two cards are not equal.
6. However, if the player mistakenly keeps two cards with expressions that are not equal, then, he/she will miss a turn as a penalty.
7. The winner is the one with the greatest number of cards.
8. When the cards in the stack are used up, the discarded cards are shuffled and the game continues.

Figure 4. Learning experience 4: Equivalent expressions

3.5 *Number sense strand 5 – Computing and counting strategies*

In Learning Experience 5, shown in Figure 5, pupils are given a mathematics version of a cloze passage with missing numerical values. Many primary school pupils are familiar with cloze passage from their English Language lessons. Pupils who lack the conceptual knowledge of whole number and money may make an attempt to fill in the blanks using guess and check. Such pupils may find this process of filling in the blanks very tedious and cumbersome. Learning Experience 5 requires the pupils to make an initial decision for a reasonable amount of Mr Lim's salary. Values for A, B, C, D and E will vary. However, Mr Lim's salary must be greater than Miss Tan's salary at first. One possible solution is: A = $25 000, B = $1000, C = $20 000, D = $2000 and E = 5. Compared to standard textbook problems that ask pupils to solve routine problems, Learning Experience 5 is an open-ended problem as it has many possible answers. Although pupils are able to compute and perform basic arithmetic operations, they have to make realistic assumptions and decisions that require reasoning in order to find the answers. Furthermore, when faced with such a word problem, pupils somehow represent its structure by identifying the quantities and the relationships between them in order to make a decision and justify it. In addition, it requires an awareness of a range of possible strategies for performing computation and an inclination to select an effective strategy. The mathematics teacher should also provide opportunities for pupils to explain their thinking so that other pupils can learn different ways of solving the problems that differ from their own.

Learning Experience 5: Whole Numbers (Primary 5)

Fill in the blanks with whole numbers so that the story makes sense.

Mr Lim's annual salary is greater than Miss Tan's. This year Mr Lim's salary is <u>A</u>. Each year he gets a raise of <u>B</u>. This year Miss Tan's salary is <u>C</u>. Each year she receives a raise of <u>D</u>. In <u>E</u> years, Mr Lim and Miss Tan will have the same salary.

Figure 5. Learning experience 5: Whole numbers

3.6 *Number sense strand 6 – Measurement of benchmarks*

Mathematics is about thinking and making sense of the world through numbers, measurements and symbols. Benchmarks are developed from personal attributes or encounters. For example, if the height of a teacher's table is 80 cm, the teacher could use this information to estimate the height of the pupils in the class. Mathematics teachers could also state a numerical situation from Learning Experience 6, as shown in Figure 6, say, what is the length (in metres) of the Benjamin Sheares Bridge? Ask pupils to make a guess. The teacher then leads a discussion of the approximations given and asks the pupils how they know. They can then find the answer from various sources, using this to develop spirit of inquiry. Once the correct answer is obtained and verified, the pupils should remember the data which can be used as a benchmark for comparing distances. Other benchmarks may include: size of water bottle, personal height and weight, dimensions of classroom, and interesting items from their home. Of course, it is not necessary to remember all the data found; the teacher has to choose appropriate ones for benchmarking and encourage students to commit these to memory. Benchmarks can also be used for fractions.

For example, the numbers 1 and $\frac{1}{2}$ could be used as benchmarks to justify that $\frac{7}{8}$ is closer to 1 than to $\frac{1}{2}$, because $\frac{7}{8}$ is closer to $\frac{8}{8}$, not to $\frac{4}{8}$, which is $\frac{1}{2}$. The final step is for the pupils to solve problems that involve real data. It is more exciting if the pupils pose their own problems that they and others solve.

Learning Experience 6: Making Approximation (Primary 5 and 6)
Match each number to its respective description.

Number	Descriptions
30	Length (in metres) of the Benjamin Sheares Bridge
5.4	Width (in kilometres) of Singapore Island
4500	Population (in millions) of Singapore as of June 2013
23	Height (in centimetres) of a teacher's table
1855	Average mass (in grams) of a school bag with books
80	Water (in litres) wasted with tap running for 5 minutes while you brush your teeth

Can you find out more facts and figures about Singapore?

Figure 6. Learning experience 6: Making approximations

These six learning experiences exemplify how teachers and pupils could benefit from learning experiences in the primary mathematics classroom. The six learning experiences are just the first step towards making number sense in the classroom meaningful where emphasis is on the process (reasoning and thinking) rather than the product (final answer).

4 Concluding Remarks

The six number sense strands highlighted above are not trivial for any pupil learning mathematics. Greeno (1991) suggested that "it may be more fruitful to view number sense as a by-product of other learning than as a goal of direct instruction" (p. 173). Similarly, Reys (1994) contended that number sense constitutes "a way of thinking that should permeate all aspects of mathematics teaching and learning" (p. 114). Number sense has to be nurtured over an extended period of time through the use of judiciously selected learning experiences by the mathematics teacher. The learning experiences presented here illustrate the six number sense strands of the framework produced by McIntosh, Reys and Reys (1992). They are in no way exhaustive. Rather this chapter provides an overview of some learning experiences that primary mathematics teachers may consider while teaching mathematics at the primary level to develop number sense.

References

Anghileri, J. (2000). *Teaching number sense.* Trowbridge, Wiltshire: Cromwell Press Ltd.

Alajmi, A. (2004). *Eighth grade Kuwaiti students' performance in recognizing reasonable answers and strategies they use to determine reasonable answers.* Unpublished doctoral dissertation, Columbia, MO: University of Missouri.

Australian Education Council (1991). *A national statement on mathematics for Australian schools.* Melbourne: Curriculum Corporation.

Berch, D. B. (2005). Making sense of number sense: Implications for children with mathematical disability. *Journal of Learning Disabilities, 38*(4), 333-339.

Burns, M. (1994). Arithmetic: The last holdout. *Phi Delta Kappan, 75*(6), 471–476.

Buxton, L. (1997). Thinking in three modes. *Mathematics Teaching, 158,* 52-53.

Doig, B., McCrae, B., & Rowe, K. (2003). *A good start to numeracy.* Melbourne, Australia: ACER.

Dunphy, E. (2007). The primary mathematics curriculum: enhancing its potential for developing young children's number sense in the early years at school. *Irish Educational Studies, 26*(1), 5-25.

Evans, D., Strnadová, I., & Wong, M. (2007). *Examining number sense: An exploratory study of students with additional learning needs.* Paper presented at the International Symposium Elementary Mathematics Teaching (SEMT '07).

Fuson, K. (1992). Research on whole number addition and subtraction. In D. A. Grouws (Ed.), *Handbook of research on mathematics teaching and learning* (pp. 243-275). New York: MacMillan.

Gersten, R., & Chard, D. (1999). Number sense: Rethinking arithmetic instruction for students with mathematical disabilities. *The Journal of Special Education, 33*(1), 18-28.

Greeno, J. G. (1991). Number sense as situated knowing in a conceptual domain. *Journal for Research in Mathematics Education, 22,* 170–218.

Griffin, S. (2004). Building number sense with number worlds: A mathematics program for young children. *Early Childhood Research Quarterly, 19*(1), 173-180.

Griffin, S., Case, R., & Siegler, R. (1994). Rightstart: Providing the central conceptual prerequisites for first formal learning of arithmetic to students at risk for school failure. In K. McGilly (Ed.), *Classroom lessons: Integrating cognitive theory and classroom practice* (pp. 24-49). Cambridge, MA: MIT.

Hiebert, J. (1986). *Conceptual and procedural knowledge: The case of mathematics.* Hillsdale, N.J.: Lawrence Erlbaum Associates.

Howden, H. (1989). Teaching number sense. *Arithmetic Teacher, 36*(6), 6-11.

Kilpatrick, J., Swafford, J., & Findell, B. (Eds.) (2001). *Adding it up: Helping children learn mathematics.* Washington, DC: National Academy Press.

Markovits, Z., & Sowder, J. T. (1994). Developing number sense: An intervention study in grade 7. *Journal for Research in Mathematics Education, 25*(1), 4-29.

McIntosh, A., Reys, B.J., & Reys, R.E. (1992). A proposed framework for examining basic number sense. *For the Learning of Mathematics, 12*(3), 2-8. 329.

McIntosh, A., Reys, B. J., Reys, R. E., Bana, J. & Farrell, B. (1997). Number sense in school mathematics: Student performance in four countries, Perth, Australia: Edith Cowan University.

Menon, R. (2004). Elementary school children's number sense. *International Journal for Mathematics Teaching and Learning.* Retrieved July 10 2013 from http://www.cimt.plymouth.ac.uk/journal/ramamenon.pdf.

Ministry of Education (2012). *Primary mathematics: Teaching and learning syllabus.* Singapore: Author.

National Council of Teachers of Mathematics (2000). *The principles and standards for school mathematics.* Reston, VA: Author.

Reys, B. J. (1994). Promoting number sense in the middle grades. *Mathematics Teaching in the Middle School, 1*(3), 114-120.

Reys, R. E., & Yang, D. C. (1998). Relationship between computational performance and number sense among sixth- and eighth-grade students in Taiwan. *Journal for Research in Mathematics Education, 29,* 225-237.

Schneider, S. B., & Thompson, C. S. (2000). Incredible equations develop number sense. *Teaching Children Mathematics, 7*(3), 165-168.

Siegler, R. S., & Booth, J. L. (2005). Development of numerical estimation: A review. In J. I. D. Campbell (Ed.), *Handbook of mathematical cognition* (pp. 197-212). New York: Psychology Press.

Sowder, J. T. (1992). Making sense of numbers in school mathematics. In G. Leinhardt, R. Putnam & R. A. Hattrup (Eds.), *Analysis of arithmetic for mathematics teaching* (pp.1-51). New Jersey: Lawrence Erlbaum Associates.

Sowder, J. T., & Klein, J. (1993). Number sense and related topics. In Owens, D. T. (Ed.), *Research Ideas for the Classroom: Middle Grades Mathematics* (pp. 41-57). National Council of Teachers of Mathematics Research Interpretation Project. New York: Simon & Schuster Macmillan.

Verschaffel, L., Greer, B., & De Corte, E. (2007). Whole number concepts and operations. In F. Lester, Jr. (Ed.), *Second handbook of research on mathematics teaching and learning* (pp. 557-628). Charlotte, NC: Information Age Publishing.

Woodward, J. & Baxter, J. (1997). *Rules and reasons: Decimal instruction for academically low achieving students.* Paper presented at the annual meeting of the American Educational Research Association, Chicago.

Wright, R. J. (1996). Problem centred mathematics in the first year of school. In J. Mulligan & M. Mitchelmore (Eds.), *Research in early number learning: An Australian perspective* (pp. 35-54). Adelaide: AAMT.

Yang, D. C. (2005). Number sense strategies used by sixth grade students in Taiwan. *Educational Studies, 31*(3), 317-333.

Yang, D. C., Hsu, C. J., & Huang, M. C. (2004). A Study of Teaching and Learning Number Sense for Sixth Grade Students in Taiwan. *International Journal of Science and Mathematics Education, 2*(3), 407-430.

Chapter 9

Learning Experiences Designed to Develop Algebraic Thinking: Lessons from the ICCAMS Project in England

Jeremy HODGEN   Dietmar KÜCHEMANN   Margaret BROWN

Algebra provides powerful tools for expressing relationships and investigating mathematical structure. It is key to success in mathematics, science, engineering and other numerate disciplines beyond school as well as in the workplace. Yet many learners do not appreciate the power and value of algebra, seeing it as a system of arbitrary rules. This may be because teaching often emphasises the procedural manipulation of symbols over a more conceptual understanding. In this chapter, we will draw on our experiences from the Increasing Competence and Confidence in Algebra and Multiplicative Structures (ICCAMS) study in order to look at ways in which learning experiences can be planned. In doing so, we will discuss how representations can be used, and the relationship between algebra and other mathematical ideas strengthened. We will also discuss how formative assessment can be used to nurture a more conceptual and reflective understanding of mathematics.

1 Introduction

Why should I learn algebra, I don't want to be a maths teacher. (A middle attaining 13 year old in England)

Algebra is a central topic within the school mathematics curriculum because of its power both within mathematics and beyond. Algebra can be used to model and predict and is thus key to science, engineering, health, economics and many other disciplines in higher education and in the workplace (Hodgen & Marks, 2013). Yet, too often we fail to communicate this power to learners who, like the 13 year old above, perceive algebra to be something that is only useful in school mathematics lessons.

The research evidence on participation in mathematics indicates that the main obstacle lies in negative learner attitudes (e.g. Matthews & Pepper, 2007; Brown, Brown, & Bibby, 2008). Most learners do not want to carry on with their mathematical studies because they believe they are not 'good at mathematics', and 'did not understand it'. They also found it 'boring' and 'unrelated to real life'. These negative attitudes apply even to many high attaining learners. Mendick (2006), for example, quotes a high attaining learner studying advanced mathematics:

What's the use of maths? ... when you graduate or when you get a job, nobody's gonna come into your office and tell you: 'Can [you] solve x square minus you know?' ... It really doesn't make sense to me. I mean it's good we're doing it. It helps you to like crack your brain, think more and you know, and all those things. But like, nobody comes [to] see you and say 'can [you] solve this?'

One can, of course, point to many contexts in which quadratics do prove useful as Budd and Sangwin (2004) have done. But we should also consider whether what we do in our mathematics classrooms could be contributing to this problem. Do we consider the difficulties that learners have with algebra sufficiently? Do we focus too much attention on algebraic manipulation and the 'rules' of algebra? Could we teach algebra in a way that conveyed its power to all learners?

In this chapter, we discuss how we addressed these problems in the Increasing Competence and Confidence in Algebra and Multiplicative Structures (ICCAMS) project. In doing so, we consider the difficulties leaners face when understanding algebra.

2 Background

ICCAMS was a 4½ year project funded by the Economic and Social Research Council in the UK. Phase 1 consisted of a survey of 11-14 years olds' understandings of algebra and multiplicative reasoning, and their attitudes to mathematics (Hodgen et al., 2010). Phase 2 was a collaborative research study with a group of teachers which aimed to improve learners' attainment and attitudes in these two areas (Brown, Hodgen, & Küchemann, 2012). Phase 3 involved a larger scale trial with a wider group of teachers and students. ICCAMS was funded as part of a wider initiative[1] aimed at increasing participation in STEM subjects in the later years of secondary school and university, a concern shared by many countries around the world including Singapore.

The Phase 1 ICCAMS survey involved a test of algebra first used in 1976 in the seminal Concepts in Secondary Mathematics and Science (CSMS) study (Hart & Johnson, 1983; Hart et al., 1981). In 2008 and 2009, the algebra test was administered to a sample of around 5000 learners aged 12-14 from schools randomly chosen to represent learners in England.

The CSMS algebra test was carefully designed over the 5-year project starting with diagnostic interviews. The original test consisted of 51 items.[2] Of these 51 items, 30 were found to perform consistently across the sample and were reported in the form of a hierarchy (Booth, 1981; Küchemann, 1981). Piloting indicated that only minor updating of language and contexts was required for the 2008/9 administration.

By using the same test that was used in the 1970s, we were able to compare how algebraic understanding had changed over the 30-year interval. Over the intervening period, there have been several large scale national initiatives that have attempted to improve mathematics teaching and learning, including learners' understanding of algebra (for a discussion of these initiatives, see Brown, 2011; Brown & Hodgen, 2013). Hence, it was a serious concern that the comparison showed that learners' understanding of algebra had fallen over time (Hodgen et al., 2010). It was in this context that we designed an approach to teaching that was intended to address learners' difficulties. However, before doing so, it is important to set out clearly exactly what we mean by 'algebraic understanding'.

2.1 *What is algebraic understanding?*

The CSMS test aims to test algebraic understanding by using "problems which were recognisably connected to the mathematics curriculum but which would require the child to use methods which were not obviously 'rules'." (Hart & Johnson, 1983, p. 2). The test items range from the basic to the sophisticated allowing broad stages of attainment in each topic to be reported, but also each item, or linked group of items, is diagnostic in order to inform teachers about one aspect of learner understanding. The focus of the test was on generalised arithmetic. Items were devised to bring out these six categories (Küchemann, 1981):

Letter evaluated, Letter not used, Letter as object, Letter as specific unknown, Letter as generalised number, and Letter as variable.

Item 5c presented the following problem to learners:

If $e + f = 8$, $e + f + g = \ldots$

Here the letters e and f could be given a value or could even be ignored; however the letter g has to be treated as at least a specific unknown which is operated upon: the item was designed to test whether learners would readily 'accept the lack of closure' (Collis, 1972) of the expression $8 + g$. Learners tend to see the expression as an instruction to do something and many are reluctant to accept that it can also be seen as an entity (in this case, a number) in its own right (Sfard, 1989). Thus, of the learners aged 13-14 tested in 1976, only 41% gave the response $8 + g$ (another 34% gave the values 12, 9 or 15 for $e + f + g$, and 3% wrote $8g$).

In question 13, learners were asked to simplify various expressions in a and b. Some of the items could also readily be solved by interpreting the letters as objects, be it as as and bs in their own right, or as a shorthand for apples and bananas, say (eg 13a: simplify $2a + 5a$; 13d: simplify $2a + 5b + a$); however, such interpretations become strained for an item like 13h (simplify $3a - b + a$), where it is difficult to make sense of subtracting a *b* (or a banana).

3 Current Approaches to Teaching in England

School algebra for 11-16 year olds in England focuses on the use of letters as *specific unknowns* rather than variables (Küchemann, Hodgen, & Brown, 2011a).[3] Also, if one looks at the more common school textbooks, the algebra is often not about anything, or at least not about anything meaningful (Hodgen, Küchemann, & Brown, 2010). Consider the example reproduced in Figure 1, which is on a page headed 'Solving problems with equations' from a homework book for learners aged 12-13.

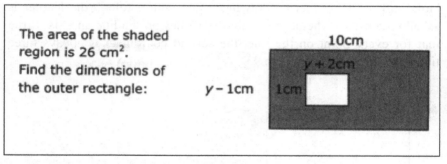

Figure 1. An example from a typical English lower secondary mathematics textbook

Here we are expected to construct and solve an equation to find a specific value of *y* and then to use this to find the dimensions of the specific rectangle that fits the given conditions. But under what circumstances (apart from when asked to practise algebraic procedures) would we want to find such an answer? And out of what kind of situation would the given expressions $y - 1$ and $y + 2$ come about? The problem becomes a lot more engaging, though not necessarily more credible, if we let *y* vary. What values of *y* 'make sense' here? What happens to the shapes of the rectangles? What is the relation between the height of the grey rectangle and the width of the white rectangle (and why have the book's authors not bothered to show this in the diagram)?

Learners most commonly meet the idea of letters as variables in the context of the Cartesian graph. Here the work is almost exclusively about straight line graphs and plotting the graph of a given function, or finding the function of a given graph. This is commonly done by focusing on the gradient and *y*-intercept on the graph, and equating this to *m* and *c* in the standard algebraic representation of the function.[4] Here, too, the work is rarely about anything. Indeed, our interviews suggest that many learners do not realise that the work is about sets of points and, moreover, points whose coordinates fit a particular relation. This contrasts strongly with the approach advocated in the best-selling *School Mathematics Project* textbook for lower secondary in the 1970s (Hodgen, Küchemann, & Brown, 2010). Here, emphasis is placed on the graph as representing a set of points all of which satisfy the relation by a consideration of the graphs with "more and more" intermediate values (see Figure 2). After "join[ing] up the points by drawing a line with a ruler", the learner is asked to consider whether the point (0, 0) lies on the line and "Is it true that for every point on the line, the second coordinate is always three times the first coordinate?" (p.93).

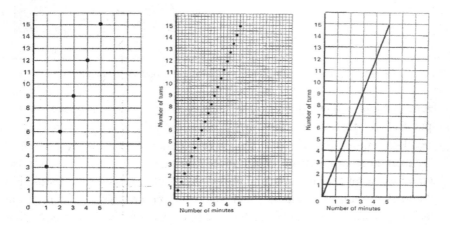

Figure 2. Three figures from a 1970s textbook

4 Our Design Principles

To counter these shortcomings, we developed a set of design principles. A key concern was to develop algebra lessons which had at least some kind of a 'realistic' context. We note that by 'realistic', we do not mean real life contexts that the learners may have encountered, but rather contexts that the learners can imagine (and indeed, in some cases, this involved a 'pure' context). In doing so, we aimed to design task that were intriguing and which provided opportunities for what Streefland (1991) refers to as the 'insightful construction of structures' (p.19). We also aimed to bring together the often fragmented activities of tabulating values, solving equations, drawing graphs, and forming and transforming algebraic expressions and relations. In addition, we drew on approaches for which there is research evidence to indicate they are effective in raising attainment (Brown et al., 2012). These included formative assessment (e.g., Black & Wiliam, 1998), connectionist teaching (e.g., Askew et al, 1997; Swan, 2006), collaborative work (e.g., Slavin, Lake, & Groff, 2009; Hattie, 2009) and the use of multiple representations (e.g., Streefland, 1993; Gravemeijer, 1999; Swan, 2008). In particular, multiple representations, such as the Cartesian graph and the double number line (see, e.g., Küchemann, Hodgen, & Brown, 2011b), are used both to help learners better understand and connect mathematical ideas and to help teachers appreciate learners' difficulties.

5 Nurturing Conceptual Understanding of Algebra

We discuss two approaches that we used, both of which link back to our earlier analysis of the teaching of algebra in England. The first takes a static textbook problem and attempts to introduce a more dynamic – and intriguing – element. The second examines how the Cartesian graph might be better introduced.

5.1 *A more dynamic approach to a textbook algebra problem*

In Figure 3, we show a task from a current English textbook, which, like the task described earlier, appears to provide little interest or intrigue.

Indeed, the triangle pictured appears to be isosceles and the height is quite visibly not twice its base.

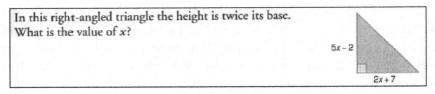

Figure 3. A static task from an English textbook

In Figure 4, we show the task as presented in an ICCAMS lesson. In one sense, the change to the task is very slight – the diagram is almost exactly the same. Yet, the additional question transforms the task to one where the learners have to think of x as a variable (or at least to consider different values for x) and then imagine what happens to the triangle as x changes. Indeed, one can start to consider whether the height *can* ever be twice the base. This might prompt further questions such as can the height ever be 10 times the base (and what would it mean for x to be negative? Is this allowed?).

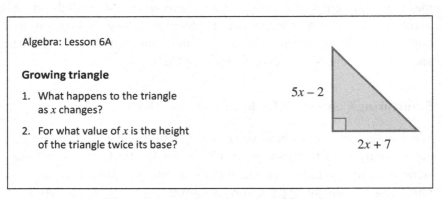

Figure 4. A more dynamic version of the task

It is a short step then to use dynamic geometry to support learners' imaginations (and to compare the original triangle to one where the height actually *is* twice the base, $4x+14$).[5]

5.2 *Making connections: Variables, tables and the cartesian graph*

One of the most interesting items on the CSMS algebra test posed the following question to learners, "Which is larger, $2n$ or $n+2$?" and asked them to explain their answer. Commonly, learners would opt for $2n$ and give a justification along the lines of 'Because it's multiply'. We were interested in whether learners would realise that the difference between the expressions varies and that there are values of n for which $2n$ can be smaller, the same, or larger than $n+2$. Very few learners demonstrated such an awareness and so we designed lessons that addressed this. One ICCAMS lesson sequence was modelled very closely on this item and began with a short whole class assessment task 'Which is larger, $3n$ or $n+3$?'. This task was designed to enable the teacher to listen to the learners' ideas and then consider (or diagnose) their (mis)-understandings prior to teaching a full lesson. (See Appendix for a reproduction of this task and the guidance given to the teacher on 'diagnosis'.) Here the context was entirely 'pure' but we designed the main task of the lesson that immediately followed this assessment to involve a 'real' context - about hiring a boat.

Algebra: Lesson 1A

Boat Hire

Olaf is spending the day at a lake.
He wants to hire a rowing boat for some of the tim.

Freya's Boat Hire charges £5 per hour.
Polly's Boat Hire charges £10 plus £1 per hour.

Whose boat should Olaf choose?

Figure 5. The Boat Hire problem

The *Boat hire* lesson started with the problem reproduced in Figure 5. The task is 'realistic' in the sense that learners can imagine such a scenario and think their way into it, even though they might never have encountered such a problem in real life, and perhaps never will. Learners found the task engaging, because they could make sense of it and because initially they came up with different conclusions which had to be resolved.

After a brief period of discussion, as a class and in small groups, the teacher is asked to record on the board the numerical data that learners come up with to support their arguments. The data are first recorded 'randomly' and then (perhaps prompted by the learners themselves) in an ordered table. Learners are used to using ordered tables, but this gives them an opportunity to see why such an ordering can be helpful.

Figure 6 shows one pair of learners' work, who, having ordered the data started noticing that the differences (-6, -2,+2, +6) form a pattern. Such analysis can prompt the question 'Are the hire costs ever the same?'. One way we suggest of pursuing this is to represent the relations algebraically (eg $5a$ and $10 + a$), which might lead learners in some classes to consider how to solve the equation $5a = 10 + a$.

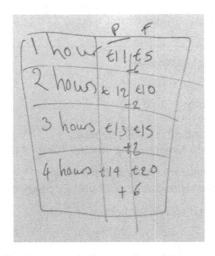

Figure 6. Two learners tabular recording of the two expressions

We also suggest putting the data on a Cartesian graph (see Figure 7). Such a representation is quite abstract (the 'picture' isn't of boats on a lake). However, because the graph is about a by-now familiar story, learners are in a good position to relate salient features of the graph to the story and the other representations. One such feature is the point where the two dotted lines cross; another might by the gradient of the lines (what does this tell us, and how is the same thing manifested in a table or algebraic expression?); or the point where a line crosses the *y*-axis (or, indeed, the *x*-axis!); or can lines meaningfully be drawn through the points (what do the intermediate points represent, and do the resulting points satisfy the relation in the table or algebraic expressions?).

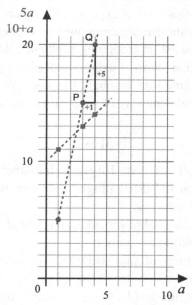

Figure 7. Using a Cartesian graph to represent 5*a* and 10+*a*

Here learners had the opportunity to see that a graph can be meaningful and useful and it sometimes lead learners to draw graphs spontaneously, eg to compare algebraic expressions. Of course, learners often had difficulties in drawing effective graphs, eg through not numbering the axes in uniform intervals, but this in itself could be a useful experience.

6 Conclusion

The ICCAMS lessons and approach was designed and trialled in the English context. In this context, the wider trial in Phase 3 of the ICCAMS study showed a significant effect; over a year the rate of learning for those learners who had experienced the lessons was double that of those who had not (Hodgen et al., 2014). We believe that the lessons, and the general approach, have wider value. Indeed, although many learners in Singapore are likely to have a better understanding of algebra than many learners in England (OECD, 2013), it is likely that they will benefit from this kind of experience.[6] In particular, there is very good evidence to indicate that algebraic understanding can be developed through an approach based on intriguing problems, making connections, promoting collaborative work, using multiple representations and diagnosing learners' (mis)-understandings (e.g., Watson, 2009). The ICCAMS lessons provide one way of supporting teachers to do this.

Notes

1. ICCAMS was part of the Targeted Initiative on Science and Mathematics Education (TISME) programme. For further information, see: tisme-scienceandmaths.org/
2. The Algebra test is available for non-commercial purposes (research and teaching) by contacting the authors.
3. An English translation of this work (Küchemann et al., 2011a) is available from the authors.
4. In England, straight line graphs are commonly referred to as $y=mx+c$.
5. A GeoGebra file of this activity is available from the authors.
6. The ICCAMS lessons are available for trialling by interested teachers and schools by contacting:
jeremy.hodgen@kcl.ac.uk or dietmar.kuchemann@kcl.ac.uk.
See also: http://iccams-maths.org

Acknowledgement

The authors are grateful to the ESRC for funding this study (Ref: RES-179-34-0001) and all the anonymous learners and teachers whose work has provided the content for this chapter.

References

Askew, M., Rhodes, V., Brown, M., Wiliam, D. & Johnson, D. (1997). *Effective teachers of numeracy*. London: King's College London.

Black, P.J., & Wiliam, D. (1998). Assessment and classroom learning. *Assessment in Education*, 5(1), 7–73.

Booth, L. R. (1981). Strategies and errors in generalized arithmetic. In Equipe de Recherche Pédagogique (Ed.), *Proc. 5th Conf. of the Int. Group for the Psychology of Mathematics Education* (Vol. 1, pp. 140-146). Grenoble: France.

Brown, M. (2011). Going back or going forward? Tensions in the formulation of a new national curriculum in mathematics. *Curriculum Journal*, 22(2), 151 - 165.

Brown, M., Brown, P., & Bibby, T. (2008). "I would rather die": Reasons given by 16 year-olds for not continuing their study of mathematics. *Research in Mathematics Education*, 10(1), 3-18.

Brown, M., & Hodgen, J. (2013). Curriculum, teachers and teaching: experiences from systemic and local curriculum change in England. In Y. Li & G. Lappan (Eds.), *Mathematics Curriculum in School Education* (pp. 377-390). Dordrecht: Springer.

Brown, M., Hodgen, J., & Küchemann, D. (2012). Changing the grade 7 curriculum in algebra and multiplicative thinking at classroom level in response to assessment data. In Sung, J.C. (Ed.), *Proceedings of the 12th International Congress on Mathematical Education (ICME-12)* (pp. 6386-6395). Seoul, Korea: International Mathematics Union.

Budd, C., & Sangwin, C. (2004). 101 uses of a quadratic equation. *Plus*. http://plus.maths.org/content/101-uses-quadratic-equation

Collis, K. F. (1972). *A study of concrete and formal reasoning in school mathematics*. (PhD thesis), University of Newcastle, New South Wales.

Gravemeijer, K. (1999). How emergent models may foster the constitution of formal mathematics. *Mathematical Thinking and Learning*, 1(2), 155-177.

Hart, K., Brown, M.L., Küchemann, D.E., Kerslake, D., Ruddock, G., & McCartney, M. (1981). *Children's understanding of mathematics: 11-16*. London: John Murray.

Hart, K. M., & Johnson, D. C. (Eds.). (1983). *Secondary school children's understanding of mathematics: A report of the mathematics component of the concepts in secondary mathematics and science programme*. London: Centre for Science Education, Chelsea College.

Hattie, J. (2009). *Visible learning: A synthesis of over 800 meta-analyses relating to achievement*. Abingdon: Routledge.

Hodgen, J., Brown, M., Küchemann, D., & Coe, R. (2010). Mathematical attainment of English secondary school students: a 30-year comparison. Paper presented at the *British Educational Research Association (BERA) Annual Conference*, University of Warwick.

Hodgen, J., Coe, R., Brown, M., & Küchemann, D. E. (2014). *Improving students' understanding of algebra and multiplicative reasoning: Did the ICCAMS intervention work?* Paper presented at the Proceedings of the Eighth British Congress of Mathematics Education (BCME8), University of Nottingham.

Hodgen, J., Küchemann, D., & Brown, M. (2010). Textbooks for the teaching of algebra in lower secondary school: are they informed by research? *Pedagogies*, 5(3), 187-201.

Hodgen, J., & Marks, R. (2013). *The employment equation: Why our young people need more maths for today's jobs*. London: The Sutton Trust.

Küchemann, D. E. (1981). *The understanding of generalised arithmetic (algebra) by secondary school children*. (PhD thesis), Chelsea College, University of London.

Küchemann, D., Hodgen, J., & Brown, M. (2011a). English school students' understanding of algebra, in the 1970s and now. *Der Mathematikunterricht*, 57(2), 41-54.

Küchemann, D. E., Hodgen, J., & Brown, M. (2011b). Using the double number line to model multiplication. In M. Pytlak, T. Rowland & T. Swoboda (Eds.), *Proceedings of the Seventh Congress of the European Society for Research in Mathematics Education (CERME7)* (pp. 326-335). Poland: University of Rzesów.

Matthews, A., & Pepper, D. (2007) *Evaluation of participation in A level mathematics: Final report*. London: QCA.

Mendick, H. (2006). *Masculinities in mathematics*. Buckingham: Open University Press.

OECD. (2013). *PISA 2012 Results: What students know and can do – Student performance in mathematics, reading and science (Volume I)*. Paris: OECD.

Sfard, A. (1989). On the dual nature of mathematical conceptions: Reflections on processes and objects as different sides of the same coin. *Educational Studies in Mathematics*, 22, (pp. 1-36).

Slavin, R. E., Lake, C., & Groff, C. (2009). Effective programs in middle and high school mathematics: A best- evidence synthesis. *Review of Educational Research*, 79(2), 839-911.

Streefland, L. (1991). *Fractions in realistic mathematics education. A paradigm of developmental research*. Dordrecht: Kluwer.

Streefland, L. (1993). Fractions: A realistic approach. In T.P. Carpenter, E. Fennema, & T.A. Romberg, (Eds.), *Rational numbers: An integration of research*. Mahwah, NJ: Lawrence Erlbaum.

Swan, M. (2006). *Collaborative learning in mathematics: A challenge to our beliefs and practices*. London: NIACE.

Swan, M. (2008). A designer speaks: Designing a multiple representation learning experience in secondary algebra. *Educational Designer*, 1(1), (http://www.educationaldesigner.org/).

Watson, A. (2009). Paper 6: Algebraic reasoning. In T. Nunes, P. Bryant, & A. Watson, (Eds.) *Key understandings in mathematics learning*. London: Nuffield Foundation.

Appendix

Guidance for *3n* or *n+3* assessment activity

Algebra: Lesson 1 STARTER

Which is larger, $3n$ or $n + 3$?

Commentary

The aim of this starter is to see what approaches students use to compare algebraic expressions.

- Do students understand the algebraic notation?
- Do they focus on the operations ('multiplication makes bigger') ?
- Do they evaluate the expressions for specific values of n ?
- Do they respond to the fact that we don't know the value of n ?
- Do they realise that the difference between the expressions might change as n varies?

Use the starter a few days before teaching the two lessons.

Chapter 10

Learning Experiences Designed to Develop Multiplicative Reasoning: Using Models to Foster Learners' Understanding

Margaret BROWN   Jeremy HODGEN   Dietmar KÜCHEMANN

Successful progress in learning mathematics depends on a sound foundation of the understanding of multiplicative structures and reasoning. This includes not only the properties and meanings of multiplication and division, but also their many links with ratio and percentage and with rational numbers - both fractions and decimals. Yet these connections take time to establish and primary teaching can sometimes emphasise facility in calculation rather than the building of conceptual connections. This chapter draws on our experiences from the Increasing Competence and Confidence in Algebra and Multiplicative Structures (ICCAMS) study in order to discuss how learning experiences can be planned to promote this kind of conceptual understanding. In doing so, the authors discuss the ways in which representations can be introduced and how formative assessment may be used.

1 Introduction

Multiplicative reasoning is a key competence for many areas of employment and everyday life (Hodgen & Marks, 2013) as well as for further mathematical study. It is however a complex conceptual field that cannot be reduced to a simple set of rules and procedures. Learners' early experiences, and understandings, of multiplication are likely to be dominated by repeated addition:

7×8 is conceived as 7 'lots of' 8 or 8+8+8+8+8+8+8.

Whilst repeated addition can be a powerful way of understanding multiplication, there is a great deal of evidence that understanding multiplication *only* as repeated addition can hinder learners' mathematical development and lead to errors and misconceptions (Anghileri, 2001). In particular, too great an emphasis on repeated addition may encourage learners to conceive of situations involving ratio or proportion as involving an additive rather than a multiplicative relationship. It is no surprise then that many researchers argue that sharing and equipartitioning provide a much stronger basis for the development of multiplicative reasoning (e.g., Confrey et al., 2009; Nunes & Bryant, 2009).

In this chapter, we begin by discussing some problems that children have in understanding multiplicative reasoning. We then consider the ways in which multiplicative reasoning can be understood and represented. Finally, we present – and discuss – two activities. In doing so, we draw on our work in the Increasing Competence and Confidence in Algebra and Multiplicative Structures (ICCAMS) study.

2 Background

ICCAMS was a 4½ year project funded by the Economic and Social Research Council in the UK.[1] Phase 1 consisted of a survey of 11-14 years olds' understandings of algebra and multiplicative reasoning, and their attitudes to mathematics (Hodgen et al., 2010). Phase 2 was a collaborative research study with a group of teachers which aimed to improve learners' attainment and attitudes in these two areas (Brown, Hodgen, & Küchemann, 2012). Phase 3 involved a larger scale trial with a wider group of teachers and learners.

The Phase 1 ICCAMS survey involved tests of decimals and ratio first used in 1976 and 1977 in the seminal Concepts in Secondary Mathematics and Science (CSMS) study (Hart & Johnson, 1983; Hart et al., 1981). In 2008 and 2009, the decimals and ratio tests were administered together with an algebra test to a sample of around 8000 learners aged 12-14 from schools randomly chosen to represent learners in England.

The CSMS tests were carefully designed over the 5-year project starting with diagnostic interviews.[2][3] The Decimals test focuses on decimals as an aspect of rational number and principally assesses two aspects of decimal number: 'measurement', and the 'multiplicative' areas of quotient and operator (Brown, Küchemann, & Hodgen, 2010). The Ratio test focuses on the use of ratio in a variety of situations with a particular focus on whether learners used multiplicative rather than additive approaches (Hart, 1980). One finding of Hart's was that there is an intermediate strategy ('rated addition') for solving ratio problems, in which, for example, instead of multiplying a number by 2.5 a learner would first double and then halve the number, and add the results together. Although both operations of 'doubling' and 'halving' have a multiplicative basis, many learners do not initially appreciate this and see them as additive operations.

In Phases 2 and 3, we developed a set of *design principles* for which there is research evidence to indicate they are effective in raising attainment (Brown, Hodgen, & Küchemann, 2012). These included *connectionist teaching* (e.g., Askew et al., 1997; Swan, 2006), *formative assessment* (e.g., Black & Wiliam, 1998), *collaborative work* (e.g., Slavin, Lake, & Groff, 2009; Hattie, 2009) and the use of *multiple representations* (e.g., Streefland, 1993; Gravemeijer, 1999; Swan, 2008). The focus of this chapter is on the last of these.

3 What is Multiplicative Reasoning?

Our approach to teaching and learning multiplicative relations uses different ways of thinking about and representing situations which involve:

- multiplication
- division
- scale factors and rates
- ratio and proportion.

These situations may involve whole numbers used to quantify discrete variables, including very large whole numbers, and positive and negative integers. They also may involve positive and negative numbers used to measure continuous variables in measurement contexts, whether rational numbers (those expressible as fractions, or as either terminating or recurring decimals) or real numbers (including non-recurring decimals like $\sqrt{2}$ or π).

The focus is on developing learners' understanding of the meaning and structure of the relationship whose general form is $c = a \times b$, and to distinguish between additive and multiplicative relationships. A solid understanding of multiplicative reasoning underlies algebra and is at the heart of the functional relation $y = kx$ (see, e.g., Davis & Renert, 2009).

4   Learners' Difficulties with Multiplicative Reasoning

Learners have a great deal of difficulty with multiplicative reasoning. So, for example, one item on the Decimals test asks learners to choose which of the two operations, multiplication ($\times$) or division ($\div$), produces the 'bigger' answer for operations involving related whole numbers and decimals (see Figure 1).

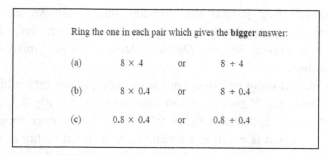

Figure 1. The 'which gives the bigger answer: × or ÷' item from the Decimals test

This item is addressed at the common misconception, often referred to as *Multiplication Makes Bigger, Division Makes Smaller*, where learners incorrectly generalise from multiplicative contexts involving whole numbers (Greer, 1994). In 2008/9, only 19% of 14 year olds in England got this item completely correct.

Understandably, the focus of a great deal of teaching is to ensure that learners know *how* to carry out multiplication and division. Yet, ensuring that learners know *when* to use multiplication or division (as well as when not to) is equally, if not more, important. Several items on the Decimals test addressed this issue. For example, learners were asked which calculation should be used to work out the price of one litre of petrol given the cost of 6.22 litres (See Figure 2).

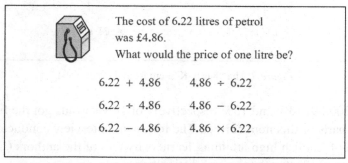

Figure 2. A 'which calculation' item from the Decimals test

In England, in 2008/9, only 17% of 14 years olds got this item correct. Many chose the incorrect response, 6.22 ÷ 4.84. This may be partly because they judged the division should always involve the division of a larger number by a smaller number, reflecting the *Multiplication Makes Bigger, Division Makes Smaller* misconception referred to above.

Contexts involving enlargement cause even greater difficulties. Item 7 of Ratio test (see Figure 3), often referred to as "Kurly K", involves relatively 'simple' multipliers [×1.5, ×2.25, ×2/3]. However, these are set in a 2D context that is much less amenable to a 'rated addition' strategy. So, whilst the numbers involved are relatively 'simple', the context is not. (For a discussion of these and related issues, see Küchemann, Hodgen, & Brown, 2011; 2014.)

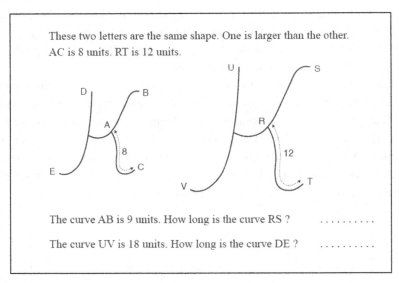

Figure 3. The "Kurly K" item from the Ratio test

In 2008/9, 13% and 15% respectively of 14 year olds got the first and second parts of this item correct. The following interview conducted with a group of English high attaining learners by two of the authors (Dietmar and Jeremy) illustrates learners' difficulties:

[The interview begins with a long silence and some whispering]

Bethan: 8 is kinda equal to 12, in the same way that 9 is equal to that [RS]

[Pause]

Dietmar: What's your next step?

Bethan: I'm still thinking ...

Zack: I don't really get it ... I was thinking ... if the 8 here [points at RS] and that's 9 [AB], you plus it to find out that? But ... is this 13? [RS]

Bethan: I'm still not sure ...

Jeremy: Why 13?

Zack: Difference is 4 ... [it's] larger by 4 ... so this [RS] should be larger (than the 9) by 4 ... if it's an accurate enlargement

Bethan: I think it's 13 and a half... cos 8 up to 12 is two-thirds... so 9 is two-thirds ... so I halved 9 which is 4.5 then add it onto 9 so you get

Jeremy: You think 13, you think 13.5 ...

Bethan: You halve it, then times it by 3 to get 3 thirds

Zack: Like 8 and 12 ... the difference 4, hey ... if you enlarge the smaller shape to get the bigger shape, wouldn't everything have to have the same ... difference?

Bethan: I'm not sure if you have to add something or times it by something ...

This, together with the item facilities, indicates that even higher attainers tend to use an addition strategy for enlargement. It is only a few higher attaining learners, like Bethan, who are able, even if tentatively, to see their way through to multiplication strategies, even when easy scale factors as here allow 'building up' or rated addition strategies. In fact in Bethan's two explanations it is possible to see a move from rated addition ('I halved...   then add it onto...') towards multiplication ('Halve it, then times it by 3').

In the next section, we consider the complexity of multiplicative reasoning by examining the different models and representations that can be used.

5 Models of, and for, Understanding Multiplicative Reasoning

Different interpretations of the *meanings* of multiplication are described in the sections below. They can be described, or represented, using models. These involve pictures and diagrams, symbols, and text to provide tools which are usable to interpret and solve problems both in the real world and in a purely mathematical context.

It is worth emphasising that none of the models described below provides a complete understanding of the nature of multiplication, or a justification for the structures of multiplication. A connected understanding of multiplicative relations involves the ability to use a range of models like these, whichever is appropriate, with any type of number, and therefore takes a great deal of time to develop. But using the models does provide opportunities for learners to discuss and share their ideas, and for teachers to highlight structural aspects of multiplication.

The models are of course not always as distinct as might be suggested, and indeed different researchers have come up with different way of categorising them. For example, Davis & Renert (2009) worked with a group of teachers who produced a list of thirteen 'realisations of multiplication' which included transformations and stretching / compressing a number line. (See also, Lamon, 2005.)

Anghileri and Johnson (1992) identified six key 'aspects' of multiplication: repeated addition or grouping, arrays and areas, scale factors and enlargements, ratio and proportion, rates, and the Cartesian product. In the following discussion, we focus on just three of these 'models'. First, we consider the strengths and weaknesses of repeated addition and the grouping model. Then, we discuss arrays and areas and scale factors and enlargements. Elsewhere we discuss the use of the double number line, and the related aspects of ratio and proportion, and rates (Küchemann, Hodgen, & Brown, 2011; 2014).

5.1 *Repeated addition and grouping*

Generally, in school and beyond, most people think of multiplication in terms of repeated addition or in terms of grouping. Thus, 7×4 can be thought of as 7 *lots of* 4 or 7(4) or as $4 + 4 \ldots + 4$. Actually there is an ambiguity about the way the expression 7×4 is interpreted in English as it can be read as (7 times) 4 i.e. 7(4), but it can also be read as 7 (times 4), or 7 multiplied by 4, which would be 4 lots of 7 or 4(7). Taking it for the moment as 7(4), this can be represented using set diagrams showing e.g. 7 packs of 4 apples (Figure 4), or as jumps on a number line (Figure 5).

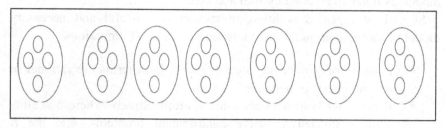

Figure 4. 7×4 modeled as 7 lots of 4

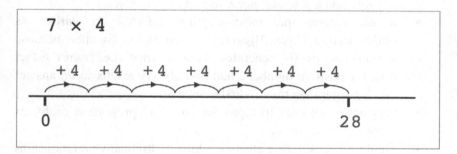

Figure 5. 7×4 modeled as 7 jumps of 4 along a number line

Generalising to the continuous case, 7 × 0.4 (e.g. for a problem about 7 glasses of 0.4 litres) can be thought of as 7 *lots of* 0.4 or as 0.4 + 0.4 ... + 0.4 or as 7(0.4). This can be shown on a number line (as can 7 × -4) but neither can be illustrated by a set diagram. However repeated addition can't easily model a product of decimals like 0.4 x7 (unless reversed) or 3.7 × 0.4 or integers like -7 × -4

Division related to the repeated addition model is more complicated than multiplication, because any given division like 28 ÷ 4 can be thought of in two ways, as involving grouping (quotition) ("How many packs of 4 apples can be made from 28 apples?") or sharing (partition)

("How many apples in each pack if 28 apples are split into 7 packs?"). Similar differences arise using the number line model between grouping ("How many steps of length 4 make 28?") and sharing ("How long are the steps if 7 steps make 28?"). Solving the sharing problem, *If I pour 7 glasses from a 2.8l bottle, how much in each glass?*, is difficult using this model, as it has to be done by trial and error.

So, whilst repeated addition/subtraction is a useful and necessary model of multiplication / division, it has a number of limitations:

- It does not generalise to cases where both terms are rational or real numbers, or negative integers.
- It does not help learners think multiplicatively where a scaling idea is needed to solve enlargement problems, e.g. the *K* enlargement on the ICCAMS tests (perhaps because one can't sensibly add a *K* to another *K* to make a larger *K*).
- It encourages the misconception referred to earlier as Multiplication Makes Bigger, Division Makes Smaller, because learners incorrectly generalise from addition (i.e. from the fact that, for positive numbers, addition always results in an answer that is bigger than the original number).
- It is not at all easy to represent ratio and proportion problems using repeated addition.
- Similarly, considering algebra, whilst multiplicative relationships involving one variable, such as 4*b*, can be straightforwardly represented on a number line, representing *ab* is more tricky as it involves a degree of imagination.
- Commutativity can be illustrated on a number line, but only to confirm that it is true: the number line does not really make clear why commutativity is **always** true.

5.2 *Arrays and areas*

Arrays (grids) and areas are very powerful tools for understanding the structure of multiplication. Whereas repeated addition represented multiplication as a unary model (one number being operated on by another), arrays and areas model multiplication as a binary operation, with the two numbers having symmetric roles. Thus 7×4 can be represented as a 7 by 4 array or a 4 by 7 array. The fact that the number of items in each of these are clearly equal provides a way of thinking about the commutative law $a \times b = b \times a$.

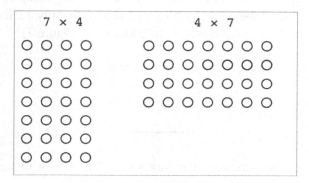

Figure 6. 7×4 and 4×7 modeled using arrays

Arrays can also be a useful way to explore factors and the associative law by discussing activities such as how many different arrays can be made from 24 dots.

The array can also be used to think about the distributive law and partitioning. For example,

$$17 \times 24 = (10 + 7)\ (20 + 4)$$
$$= (10 \times 20) + (7 \times 20) + (10 \times 4) + (7 \times 4)$$

Whereas arrays involve discrete contexts, areas use what is virtually the same model but with continuous contexts and numbers expressed as decimals or fractions.

For division problems, arrays and areas can be very useful in helping learners to understand division as the inverse of multiplication, leaving behind ideas of repeated addition and subtraction. Hence, we can solve $72 \div 12$ by asking what number do you multiply 12 by to get 72. Similarly, we can model $7.2 \div 0.6$ using a rectangle of side 0.6 and area 7.2.

It is important to note that areas are not absolutely straightforward as learners may not fully understand what area is (i.e. finding the number of unit squares that cover a region completely). It is likely that they will be developing their understandings of area and multiplication alongside each other. So, on the ICCAMS Algebra test (see Chapter 9 of this book), when asked the area of a 5 by e+2 rectangle, only 13% of 14 year olds in England answered correctly in 2008/9 (see Figure 7).

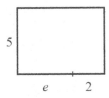

Figure 7. A 'What is the area of this shape' [5 × (e+2)] item from the Algebra test

Areas also generalise to algebra in the use of areas as models for the multiplication of variables. The textbooks tend to assume that this representation is transparent. But unless learners have some experience of using areas to represent multiplication, the multiplicative relationships that are at the heart of the algebra are not at all clear. Hence, learners may be learning *how* to work out problems involving symbols and shapes rather than using area as a tool / representation of multiplication to develop their understanding of algebra.

As with arrays, an important feature of areas is that the commutative law is powerfully represented, and the symmetry of the two numbers or variables involved. This in turn can help learners to abstract from particular models and contexts by developing an understanding of how they can use what works best for the numbers (and to develop algebraic thinking).

5.3 *Scale factors and enlargements*

The idea of a multiplying factor is really a generalisation of the repeated addition model where 7(4) could be thought of as 7 lots of 4 (e.g. 7 packs of 4 apples), and where 7 could therefore really be thought of as a multiplying factor operating on the number 4, with ×7 as a unary operation.

Although the repeated addition model required the multiplying factor to be a whole number, the extension of the notion of a multiplying factor to a scale factor acting on a length on the number line allows a generalisation to a rational number as multiplier. Thus 3.7×0.4 can be thought of as enlarging (stretching) a number line by a scale factor of 3.7, so that a length of 0.4 becomes $3.7 \times 0.4 = 1.48$. Similarly 0.4×3.7 becomes an 'enlargement' of the number line by a scale factor of 0.4, which is less than 1 and hence a shrinking.

We can symbolise this enlargement relationship as $y = m \times x$, since the scale factor, m, is constant and the initial length on the number line, x, varies. This is the case therefore that has the closest relationship with the linear relationship $y = mx + c$. It also means that it makes sense to represent an enlargement on a graph as $y = mx$.

In 2 dimensions we could 'stretch' each dimension by a different scale factor but in enlargement problems relating to similar figures we have a single scale factor, as in the K enlargement, so we are interested in learners understanding that the scale factor is constant for all corresponding elements of each shape. So, each length, x, in the smaller K is enlarged by a scale factor of $^3/_2$, thus becoming $(^3/_2 \times x)$ in the enlarged K. Again the relation can be graphed as $y=3x/2$ or $y=1.5x$.

There are two forms of division relating to scaling problems, one of which reflects grouping and one sharing. For the former, we can use two corresponding lengths to find out the scale factor, which corresponds to dividing 28 apples by 4 apples to find the multiplying factor of 7, which was a grouping problem. While it is in the discrete case solvable by repeated addition or subtraction, in the context of enlargement this doesn't always make any sense (e.g. how many small curly sides can you put together to make a curly side 3.7 times as long?).

The 'sharing' model of division ($28 \div 7 = 4$) corresponds to the other enlargement division problem where the resulting length is known and the scale factor, and we have to work out what was the length we started with.

6   Using Models to Develop Multiplicative Reasoning

In this section we will briefly describe tasks used in ICCAMS lessons aimed at developing models for multiplication and ratio.[4]

6.1 Whole number multiplication models

Since one focus of the design of ICCAMS teaching was on formative assessment (Hodgen & Wiliam, 2006), an early lesson in the sequence is intended to allow the teacher to assess where learners are starting from in their understanding of the different verbal and visual models for multiplication.

Models and stories

Here is an expression involving 12 and 3:

Think of   a. some ways of saying "12 × 3"
           b. some ways of calculating 12 × 3
           c. some diagrams that fit the expression
           d. some stories that fit the expression.

$$12 \times 3$$

Figure 8. An assessment task: Models and stories for 12 × 3.

The task in Figure 8 is designed to elicit a range of models for multiplication and, thus, to enable class discussion of different models proposed by learners, as well as suggesting to the teacher how much work is needed to build familiarity with the models before proceeding to the introduction of more sophisticated work on scale factors and ratio.

6.2 Using areas to model multiplication with rational numbers

A key issue is to extend the area model from multiplication with whole numbers to multiplication with rational numbers. The task illustrated in Figure 9, begins by considering 3.2 × 2.6 as 3.2 rows of 2.6 unit squares. Learners can estimate the number of squares as well as discuss how many unit squares there are in the "thin" column to the right, the "short" row at the bottom, or the small rectangle on the right at the bottom.

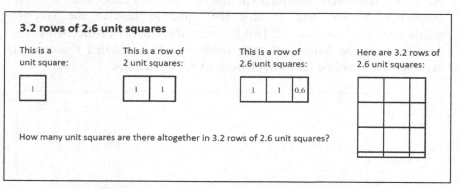

Figure 9. The task involving 3.2 × 2.6 as 3.2 rows of 2.6 unit squares

This meaning is a little strained for fractional numbers. Hence, the lesson moves on to a task (involving 5.3 × 2.6) in which the area is thought of as distances along orthogonal number lines whose product is an area (whilst the area is still conceived of in terms of unit squares).

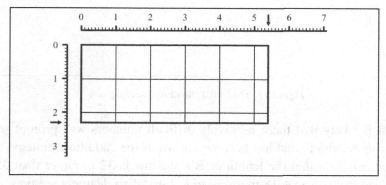

Figure 10. The task involving 5.4 × 2.3 using orthogonal number lines

6.2 Modeling relationships involving scaling

Here we present part of a sequence of related scaling tasks set in the context of shadows, where the tasks are used to diagnose learners' understandings in order to adapt the task adapted to learners' needs. The numbers in Figure 11 have deliberately been chosen so that the (functional) relationship between the length of a post and its shadow involves a fractional multiplier (in this case ×2 1/2) rather than a whole number multiplier, and so that the same applies to the (scalar) relationship (in this case ×2 1/6) between the lengths of the two poles (and between the lengths of their shadows). The fact that the context involves ratio may be far from obvious to some students.

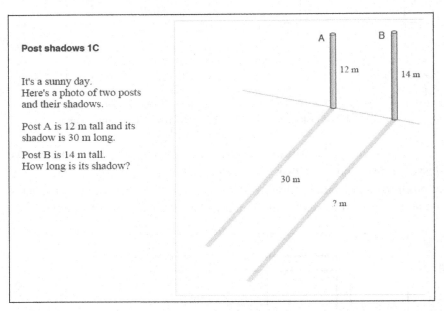

Post shadows 1C

It's a sunny day.
Here's a photo of two posts and their shadows.

Post A is 12 m tall and its shadow is 30 m long.

Post B is 14 m tall.
How long is its shadow?

A
B

12 m
14 m

30 m

? m

Figure 11. The initial 'Shadows' scaling task[5]

It is likely that these relatively difficult numbers will prompt some students to adopt what has become known as the "addition strategy" and hence conclude that the length of B's shadow is 32 m rather than 35 m (derived from 14 m + 18 m, or 30 m + 2 m). If so, learners are presented with the task in Figure 12. Here, post B has been partitioned into two sections, one of which is the same length as post A and which hence has a shadow of 30 m. The other section is just 2 m long. The task now is to find the length of the shadow of the 2 m section. The intention here is to provoke a conflict for those students who attempt to apply the addition strategy again, as this leads to a shadow length of 20 m (12 m + 18 m, or 30 m – 10 m), which is far larger than the actual length of 5 m.

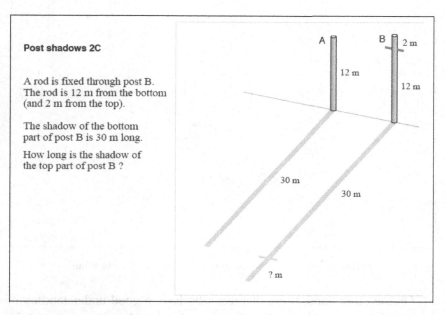

Post shadows 2C

A rod is fixed through post B. The rod is 12 m from the bottom (and 2 m from the top).

The shadow of the bottom part of post B is 30 m long.

How long is the shadow of the top part of post B ?

Figure 12. The 'Shadows' scaling task adapted for learners using the addition strategy

If most learners come up with the correct answer of 35 m on Figure 11 (and thus don't use the addition strategy), they may be presented with the task in Figure 13. In this task the sun has moved to a slightly lower elevation, causing the shadow of the 12 m post to lengthen by 1 m. The (functional) relationship between post length and shadow length is now numerically even more complex (×2 7/12 rather than ×2 1/2), which may lead some students to resort to the addition strategy and hence give an answer of 36 m (35 m + 1 m) or 33 m (31 m + 2 m, or 14 m + 19 m) for the new shadow of post B, rather than 36 1/6 m.

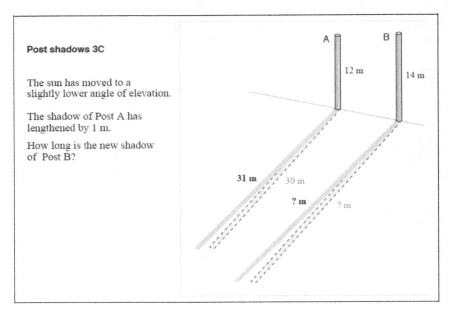

Figure 13. The 'Shadows' scaling task adapted for those using

On the other hand, those learners who have a good understanding of the multiplicative relationship between the length of a post and its shadow may have no markedly greater difficulty in deciding how to find the relationship (e.g. divide 31 by 12) and how to apply it to find the new shadow of post B (e.g., $14 \times (31 \div 12)$) - though some may need/want to use a calculator to carry this out.

7 Conclusion

The ICCAMS lessons and approach was designed and trialled in the English context. In this context, the wider trial in Phase 3 of the ICCAMS study showed a significant effect: over a year the rate of learning for those who had experienced the lessons was double that of those who had not (Hodgen et al., 2014). We have focused here on three models for developing understanding of multiplicative reasoning and shown how the repeated addition model, whilst important, is not sufficient for developing multiplicative reasoning as it breaks down for multiplicative situations such as scaling, and when we shift from the natural numbers to integers and rational numbers

Notes

1. ICCAMS was part of the Targeted Initiative on Science and Mathematics Education (TISME) programme. For further information, see: tisme-scienceandmaths.org/
2. The Decimals and Ratio tests are available for non-commercial purposes (research and teaching) by contacting the authors.
3. A few additional items from the original Fractions test were added to the Ratio test in order to assess learners' understanding in this area. Piloting indicated that only minor updating of language and contexts was required for the 2008/9 administration.
4. The ICCAMS lessons are available for trialling by interested teachers and schools by contacting:
jeremy.hodgen@kcl.ac.uk or dietmar.kuchemann@kcl.ac.uk.
See also: http://iccams-maths.org
5. Note that this picture is drawn in the Northern hemisphere.

Acknowledgement

The authors are grateful to the ESRC for funding this study (Ref: RES-179-34-0001) and all the anonymous learners and teachers whose work has provided the content for this chapter.

References

Anghileri, J. (2001). British research on mental and written calculation methods for multiplication and division. In M. Askew & M. Brown (Eds.), *Teaching and learning primary numeracy: Policy, practice and effectiveness. A review of British research for the British Educational Research Association in conjunction with the British Society for Research into Learning of Mathematics* (pp. 22-27). Southwell, Notts: British Educational Research Association (BERA).

Anghileri, J., & Johnson, D. C. (1992). Arithmetic operations on whole numbers: multiplication and division. In T. R. Post (Ed.), *Teaching mathematics in grades K-8: Research based methods* (Second ed., pp. 157-200). Boston, MA: Allyn and Bacon.

Askew, M., Rhodes, V., Brown, M., Wiliam, D., & Johnson, D. (1997). *Effective teachers of numeracy*. London: King's College London.

Black, P.J., & Wiliam, D. (1998). Assessment and classroom learning. *Assessment in Education*, 5(1), 7–73.

Brown, M., Hodgen, J., & Küchemann, D. (2012). Changing the grade 7 curriculum in algebra and multiplicative thinking at classroom level in response to assessment data. In Sung, J.C. (Ed.), *Proceedings of the 12th International Congress on Mathematical Education (ICME-12)* (pp. 6386-6395). Seoul, Korea: International Mathematics Union.

Brown, M., Küchemann, D. E., & Hodgen, J. (2010). The struggle to achieve multiplicative reasoning 11-14. In M. Joubert & P. Andrews (Eds.), *Proceedings of the Seventh British Congress of Mathematics Education (BCME7)* (Vol. 30, pp. 49-56). University of Manchester: BSRLM.

Confrey, J., Maloney, A. P., Nguyen, K. H., Mojica, G., & Myers, M. (2009). Equipartitioning / splitting as a foundation of rational number reasoning using learning trajectories. In M. Tzekaki, M. Kaldrimidou & H. Sakonidis (Eds.), *Proceedings of the 33rd Conference of the International Group for the Psychology of Mathematics Education* (pp. 345-352). Thessaloniki, Greece: PME.

Davis, B., & Renert, M. (2009). Mathematics-for-teaching as shared dynamic participation. *For the Learning of Mathematics*, 29(3), 37-43.

Gravemeijer, K. (1999) How emergent models may foster the constitution of formal mathematics. *Mathematical Thinking and Learning*, 1(2), 155-177.

Greer, B. (1994). Extending the meaning of multiplication and division. In G. Harel & J. Confrey (Eds.), *The development of multiplicative reasoning in the learning of mathematics* (pp. 61-85): SUNY Press.

Hart, K.M. (1980). *Secondary school children's understanding of ratio and proportion* (Ph D thesis). Chelsea College, University of London.

Hart, K., M. L. Brown, D. E. Küchemann, D. Kerslake, G. Ruddock, and M. McCartney, (Eds). (1981). *Children's understanding of mathematics: 11-16.* London: John Murray.

Hart, K. M., & Johnson, D. C. (Eds.). (1983). *Secondary school children's understanding of mathematics: A report of the mathematics component of the concepts in secondary mathematics and science programme.* London: Centre for Science Education, Chelsea College.

Hattie, J. (2009) *Visible learning: A synthesis of over 800 meta-analyses relating to achievement.* Abingdon: Routledge.

Hodgen, J., Brown, M., Küchemann, D., & Coe, R. (2010). Mathematical attainment of English secondary school students: a 30-year comparison. Paper presented at the *British Educational Research Association (BERA) Annual Conference*, University of Warwick.

Hodgen, J., Coe, R., Brown, M., & Küchemann, D. E. (2014). *Improving students' understanding of algebra and multiplicative reasoning: Did the ICCAMS intervention work?* Paper presented at the Proceedings of the Eighth British Congress of Mathematics Education (BCME8), University of Nottingham.

Hodgen, J., & Marks, R. (2013). *The employment equation: Why our young people need more maths for today's jobs.* London: The Sutton Trust.

Hodgen, J., & Wiliam, D. (2006). *Mathematics inside the black box.* London: NFER-Nelson.

Küchemann, D. E., Hodgen, J., & Brown, M. (2014). *The use of alternative double number lines as models of ratio tasks and as models for ratio relations and scaling.* Paper presented at the Proceedings of the Eighth British Congress of Mathematics Education (BCME8).

Küchemann, D. E., Hodgen, J., & Brown, M. (2011). Using the double number line to model multiplication. In M. Pytlak, T. Rowland & T. Swoboda (Eds.), *Proceedings of the Seventh Congress of the European Society for Research in Mathematics Education (CERME7)* (pp. 326-335). Poland: University of Rzesów.

Lamon, S. J. (2005). *Teaching fractions and ratios for understanding: Essential content knowledge and instructional strategies for teachers.* Mahwah, NJ: Lawrence Erlbaum.

Nunes, T., & Bryant, P. (2009). Paper 3: Understanding rational numbers and intensive quantities. In T. Nunes, P. Bryant & A. Watson (Eds.) *Key understandings in mathematics learning.* London: Nuffield Foundation. Available from www.nuffieldfoundation.org, accessed 9 October 2009.

Slavin, R.E., Lake, C., & Groff, C. (2009). Effective programs in middle and high school mathematics: A best- evidence synthesis. *Review of Educational Research*, 79(2), 839-911.

Streefland, L. (1993). Fractions: A realistic approach. In Carpenter, T.P., Fennema, E. & Romberg, T.A. (Eds.), *Rational numbers: An integration of research.* Mahwah, NJ: Lawrence Erlbaum.

Swan, M. (2006). *Collaborative learning in mathematics: A challenge to our beliefs and practices.* London: NIACE.

Swan, M. (2008). A designer speaks: Designing a multiple representation learning experience in secondary algebra. *Educational Designer,* 1(1), (http://www.educationaldesigner.org/).

Chapter 11

Learning Mathematical Induction through Experiencing Authentic Problem Solving

TAY Eng Guan   TOH Pee Choon

The topic of mathematical induction is anomalous in the mathematics syllabus in that it is not within any field of mathematics but rather it is a technique. This 'abnormality' would perhaps suggest a different way of teaching. This paper sets the pedagogy of the technique of mathematical induction within its natural environment of problem solving where a problem is explored, a conjecture is made, and an attempt to prove the conjecture using some techniques is made on the basis of the earlier exploration.

1 Introduction

Standard 3 (Mathematics as Reasoning) of the National Council of Teachers of Mathematics (NCTM) Curriculum and Evaluation Standards for School Mathematics (1989, p. 143) includes the following:
"and so that, in addition, college-intending students can—

- construct proofs for mathematical assertions, including indirect proofs and proofs by mathematical induction."

Elaborating on the three goals of the Standard, the NCTM explains that "[a] third goal, also a departure from the existing curriculum for college-intending students, is to give increased attention to proof by mathematical induction, the most prominent proof technique in discrete mathematics." Thus, mathematical induction (abbreviated as MI from now) is seen by NCTM as a crucial technique to be learnt by college-intending students.

The technique of MI is also included in the Singapore mathematics syllabus for the University of Cambridge Local Examinations Syndicate at Advanced Level for many years. However, a quick survey of the 1998, 2006 and 2013 syllabuses of the University of Cambridge Local Examinations Syndicate (1998, 2006, 2013) shows that the scope and depth of study of MI has been drastically reduced from a topic of study in both Mathematics Syllabus C and Further Mathematics Syllabus C to just one technique of proof within the topic of Summation of Series in the new H2 Mathematics syllabus. Table 1 details the content descriptions relevant to MI.

Table 1
MI related content in the syllabus

Subject (Year)	Topic	Content
Mathematics Syllabus C (1998)	The method of induction	Problems may involve the summation of finite series
Further Mathematics Syllabus C (1998)	Further MI	Problems set may involve divisibility tests and inequalities
Mathematics A Level (2006)	MI	Understand the steps needed to carry out a proof by the method of induction; use the method of induction to establish a given result e.g. the sum of a finite series, or the form of an nth derivative.
Further Mathematics A Level (2006)	Further MI	Use the method of induction to establish a given result (questions set may involve divisibility tests and inequalities, for example); recognise situations where conjecture based on a limited trial followed by inductive proof is a useful strategy, and carry this out in simple cases e.g. to find the nth derivative of xe^x.
H2 Mathematics (2013)	Summation of series	Proof by the method of mathematical induction

Researchers such as Brumfiel (1974), Avital and Libeskind (1978), Dubinsky (1986, 1989), Baker (1996) and Foret (1998) have examined the learning of MI and concluded that it was a very difficult concept for high school students to master. Most research attempted to find out student misconceptions and learning difficulties, for example, a local study by Chow (2002, 2003) sought only to find out the level of mastery of MI among junior college students, and the learning difficulties that they experienced when studying MI. Chow (2002) concluded that errors are starting points in students learning. This claim is in line with what Borasi (1985) found: "[E]rrors should be exploited as springboards for the learning of mathematics" (as cited in Chow, 2002, p. 76).

Yet, it is feared that the problem goes deeper than just a collection of errors to be remedied piecemeal. Dubinsky (1986) commented on the deep malaise of MI instruction and use: "If you question students—even those who have had several mathematics courses—although almost all of them would have heard of induction, not many of them will be able to say anything intelligent about what it is, much less actually use it to solve a problem" (p. 305).

Harel (2002) instead suggested that the learning of MI would be more effective if MI were developed as a proof scheme. His instructional treatment of MI first took into account two deficiencies in "standard instructional treatment of MI" (p. 194): (1) MI was introduced abruptly, that is, students did not see why it was needed nor its connection to previous problem solving heuristics; and (2) the type and order of the problems used, in that they focused almost exclusively on certain types of problems which students learnt to solve by blindly following the two MI steps of verifying the base case and manipulating to obtain the induction step.

Harel (2002) then further described a new instructional treatment of MI that was structured in three phases to help students develop the understanding of MI through suitably chosen problems. A key objective of this new treatment was to induce students to focus on process pattern generalization rather than on result pattern generalization. The following example may help to illustrate the distinction: Given the sequence of the first five terms of the square numbers represented by a sequence of dots which form a square, a person who focuses on result pattern generalization will count the number of dots in the first five figures as 1, $4 = 2^2$, $9 = 3^2$, $16 = 4^2$, $25 = 5^2$ and conclude that the next number must be $6^2 = 36$. On the other hand, the person who focuses on process pattern generalization will see that the number of dots added to make the next square will be twice the length of the larger square less one. Thus, he will obtain $25 + (2 \times 6 - 1) = 36$. This will form a natural basis for inductive reasoning.

The three phases in Harel (2002) were (i) quasi-induction as an internalized (i.e. autonomous and often spontaneous part of problem solving repertoire) process pattern generalization; (ii) quasi-induction as an interiorized (i.e. ability to represent it conceptually as a method of proof) process pattern generalization; (iii) MI as an abstraction of quasi-induction. It was reported that the instruction was successful for undergraduate students: "The students seemed to have easily assimilated the principle of MI into their scheme of quasi-induction: About 75% of the problems assigned were solved correctly and in terms of the principle of MI; the rest were either solved by quasi-induction or by other means" (Harel, 2002, p. 203).

This chapter is an attempt along the line of Harel to move away from the teaching of MI as a series of steps to take and within a restricted number of types of problems. It leverages on the curriculum guidelines of the Singapore Ministry of Education (see for example, MOE, 2012) which states the centrality of problem solving in the curriculum and introduces the concept of "learning experiences" for the purpose of acquiring skills.

2 Problem Solving and Learning Experience

The conceptualization of the Singapore mathematics curriculum is represented by the "pentagonal framework" (MOE, 2012, p.14) of Figure 1, which has been the guiding framework for curriculum planning and revision since the late 1980s. We can see that the components of the pentagon—attitude, skills, concepts, processes, and metacognition—are supportive of the heart of the curriculum: mathematical problem solving.

"The central focus of the framework is mathematical problem solving, that is, using mathematics to solve problems."

Beliefs
Interest
Appreciation
Confidence
Perseverance

Monitoring of one's own thinking
Self-regulation of learning

Numerical calculation
Algebraic manipulation
Spatial visualisation
Data analysis
Measurement
Use of mathematical tools
Estimation

Reasoning, communication and connections
Applications and modeling
Thinking skills and heuristics

Numerical
Algebraic
Geometric
Statistical
Probabilistic
Analytical

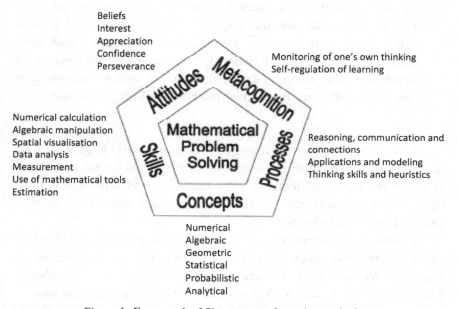

Figure 1. Framework of Singapore mathematics curriculum

We have a narrower focus of 'problem solving' which is in line with much of the literature (eg., Kroll & Miller, 1993, Krulik & Rudnik, 1980, Lester, 1994). We define problem solving as an activity that involves problem solvers viewing problems as unfamiliar or non-routine which requires them to develop productive strategies to solve the problem. Problem solving is to us a different activity from repetitive practices of standard procedures to complete familiar mathematical exercise questions. Certainly, this narrower focus of problem solving is an essential part of the larger perspective of "using mathematics to solve problems".

In this most recent of curriculum revisions by the Ministry of Education (MOE, 2012), the phrase "learning experience" is used to encourage teachers to create opportunities for the development of the 'process skills':

> Learning mathematics is more than just learning concepts and skills. Equally important are the cognitive and metacognitive process skills. These processes are learnt through carefully constructed learning experiences. For example, to encourage students to be inquisitive, the learning experiences must include opportunities where students learn to discover mathematical results on their own. (p. 20)

From the *Longman Dictionary of Contemporary English* (1978), 'experience' is explained as "something that happens to one and has an effect on the mind and feelings." The implication of this description is that a 'learning experience' needs to leave an impression on the memory and the heart for it to be successful.

Combining these two motivations of learning problem solving skills and a learning experience that is impactful, we presented a workshop on Teaching Mathematical Induction through Experiencing Authentic Problem Solving in the 2013 Association of Mathematics Educators – Singapore Mathematical Society Conference. Thirty-six teachers, mostly from the junior colleges and the integrated programme schools, attended the workshop. The next section describes the main activities and teacher responses of the workshop.

3 Teaching Mathematical Induction through Experiencing Authentic Problem Solving

The workshop began with an exposition on the anomaly of MI in the Advanced Level syllabus because MI is not within a particular field of mathematics but rather it is a technique. This anomaly would perhaps suggest a different way of teaching. It was then stated that the workshop would set the pedagogy of the technique of MI within its natural environment of problem solving where (i) a problem is explored, (ii) a conjecture is made, and (iii) an attempt to prove the conjecture using some techniques is made on the basis of the earlier exploration.

One of the authors, who was the facilitator (henceforth referred to as the facilitator), then explained that a learning experience ought to be personal, memorable and understood.

3.1 *Pólya's problem solving model and the 'natural' use of MI*

The first activity was based on Problem 1 (see Cai & Brook, 2006, p. 43) below. The purpose of the activity was to let the participants realize that by exploring, their students would make conjectures in an authentic manner. Also, by seeing where the conjecture came from, ideas of how to prove the conjecture, which may be by using MI, would be formed.

Problem 1 (Pyramid of odd numbers)
Find and prove some number patterns from the pyramid of consecutive odd numbers below where the n-th row contains n odd numbers.

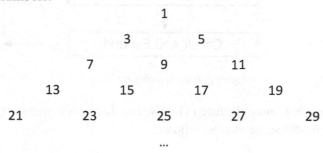

The participants were engaged in the activity and a number of conjectures were obtained, such as:

1. The first number in the n-th row is $1 + 2 + 4 + 6 + \ldots + 2(n-1)$.
2. The last number in the n-th row is $1 + 4 + 6 + 8 + \ldots + 2n$.
3. The number in the centre of the $(2n-1)$-th row is $(2n-1)^2$.
4. The sum of the numbers in the n-th row is n^3.

For the first two conjectures, some participants provided 'proofs' by summing up the series up to the n-th term, not realizing that they needed first to prove that the conjecture was true. At this point, the facilitator explained that "pattern is not proof" (see Toh et al., 2011, p. 128). There was a need to prove that indeed the n-th terms of the series were $2(n-1)$ and $2n$ respectively. Thus, in solving a problem, one ought to be clear about what is required of the problem and so one ought to understand the problem. The facilitator thus asked the participants to recall Pólya's famous problem solving model (see *Figure 2* and note that we replaced "Look Back" with "Check and Expand").

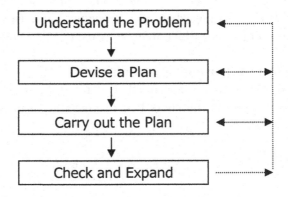

Figure 2. Pólya's problem solving model

Focusing first on conjecture (1) only, the discussion was now framed by Pólya's model somewhat as follows:

UP1: Understand the Problem (First Attempt)
We are required to look for a pattern and we found one in the first numbers of each row: 1, 3, 7, 13 and 21. We see that the 'add-on' for the terms are +2, +4, +6, +8.

DP1: Devise a Plan (First Plan)
We conjecture that the n-th term is $1 + 2 + 4 + 6 + \ldots + 2(n-1)$. We will prove that the n-th term to be added in the series is $2(n-1)$.

CP1: Carry out the Plan (First Plan)
Note that the number of terms in the $(n-1)$-th row is $n-1$. Since the terms are consecutive odd numbers, the difference between the first term of the $(n-1)$-th row and the first term of the n-th row is $2(n-1)$.

DP2: Devise a Plan (Second Plan)
We will use the Arithmetic Progression formula to find $1 + 2 + 4 + 6 + \ldots + 2(n-1)$.

CP2: Carry out the Plan (Second Plan)
Thus the n-th term $= 1 + 2 + 4 + 6 + \ldots + 2(n-1)$
$= 1 + (n-1)(2 + 2(n-1))/2 = 1 + n(n-1)$.

C/E2: Check and Expand (Second Plan)
Check formula holds for $n = 1, 2, 3$. Unlike Harel (2002), we assumed that the MI method had been taught earlier to the students in a didactic manner as this would be most efficient given the allocated time for MI instruction in the curriculum. So, we suggested that MI can be used to give a rigorous solution.

DP3
Write a rigorous proof by using MI.

CP3
For positive integers n, let P(n) be the statement: The first term in the n-th row is $1 + n(n-1)$. When $n = 1$, first term in the n-th row $= 1 = 1 + 1(1-1)$. Thus, P(1) is true. Suppose P(k) is true for some $k \geq 1$, i.e. the first term in the k-th row is $1 + k(k-1)$. There are k odd numbers in the k-th row so the difference between the first term in the $(k+1)$-th row and the first term in the k-th row is $2k$. Hence, the first term in the $(k+1)$-th row $= 1 + k(k-1) + 2k$ (by induction hypothesis) $= 1 + (k+1)k$. Since P(1) is true, and P(k) implies P($k+1$), by MI, P(n) is true for all $n \geq 1$.

We note that the idea for the inductive step came from the exploration afforded by using Pólya's problem solving model. Students working on this one problem would have a number of sub-problems (patterns) arising from it and conjectures for these would be natural reasons to use MI. As a further example, the inductive step for Conjecture (3) would be based on the observation that the number of odd numbers from the centre number of the $(2n - 1)$-th row (not inclusive) to the centre number of the $(2n + 1)$-th row (inclusive) is $(n - 1) + 2n + (n + 1)$.

3.2 *Judging the validity of an MI argument*

The participants were then given a MI 'proof' for them to judge its validity. The problem (Problem 2) was an adaptation of a standard undergraduate problem in Logic.

Problem 2 (Smart as Einstein)
Here is a proof by mathematical induction that you are as smart as Einstein!

Theorem: I am as smart as Einstein.

Proof: Let $P(n)$ be the statement: All n persons in a group containing n persons have the same IQ.

$P(1)$ is obviously true.

Suppose $P(k)$ is true for some positive integer k.

Take a group X of $k + 1$ persons.
Remove a person A from the group, leaving behind a group X' of k persons. By the induction hypothesis, all the k persons in X' have the same IQ.

Remove a person B from X' and put back the first removed person A, thus forming a set X'' of k persons. By the induction hypothesis, all the k persons in X'' have the same IQ.

Finally, put back the second removed person B to re-form the group X with $k + 1$ persons, all of whom have the same IQ. Thus if $P(k)$ is true, then $P(k + 1)$ is also true.

Since $P(1)$ is true and $P(k) \Rightarrow P(k + 1)$, by mathematical induction, $P(n)$ is true for all positive integers n, i.e. all n persons in a group containing n persons have the same IQ. Finally, put me in a group with Einstein, and I will be as smart as he is!

Most of the participants were not able to detect the flaw in the argument. Some thought that the flaw lay in the base case $n = 1$. Though this was partly true (the base cases should have been $n = 1, 2$), they pointed out wrongly that $P(1)$ was false. A participant was able to come to the front and show on the whiteboard that the inductive step was flawed because it applied only for $k \geq 2$ by showing specifically what happened when $k = 1$, i.e. that there is no person then to 'link' the groups X' and X''.

Another participant recalled to his fellow participant that the problem was similar to one encountered long ago in their undergraduate class but was never resolved till then. The facilitator then impressed that it was important for students to have a chance to judge when an argument was wrong to deepen their understanding of the MI technique.

3.3 *Pólya's problem solving model and aiming for sub-goals in MI*

The final problem in the workshop was the Tower Function problem. The intention was for participants to have another chance at using Pólya's problem solving model and in addition, focus on the heuristic of 'aiming for sub-goals' in the use of the MI technique. The problem is presented below.

Problem 3 (Tower function)

Let $a^{(n)}$ denote $a^{a^{a^{\cdot^{\cdot^{\cdot a}}}}}$, a tower of n terms of a. Find the smallest integer m such that $9^{(n)} < 3^{(m)}$.

The participants were given time to solve the problem on their own and the subsequent discussion led by the facilitator would be somewhat as follows.

UP1
We use numbers to familiarize ourselves with the new function. Thus, $3^{(2)} = 3^3 = 27$; $3^{(3)} = 3^{3^3} = 3^{27} = 7625597484987$. Note that $3^{3^3} \neq 27^3 = 19683$.

DP1
List out $3^{(m)}$ and $9^{(n)}$ for $m = 1, 2, 3$ and $n = 1, 2, 3$, and compare the values to have an idea of how much bigger m must be compared to n.

CP1
From *UP1*, we have $3^{(1)} = 3$; $3^{(2)} = 3^3 = 27$; $3^{(3)} = 3^{3^3} = 3^{27} = 7625597484987$. $9^{(1)} = 9$; $9^{(2)} = 9^9 = 387420489$; $9^{(3)} = 9^{9^9}$, which is too large to be computed by a calculator. But we managed to see that $3^{(2)} > 9^{(1)}$ and $3^{(3)} > 9^{(2)}$.

UP2
We conjecture that $m = n + 1$. We restate the problem to: Show that $3^{(n+1)} > 9^{(n)}$ for all positive integers n.

DP2
We want to find out without using the calculator why $3^{(n+1)} > 9^{(n)}$ for $n = 2$ and 3, by manipulating the expressions to obtain the same base.

CP2
$3^{(3)} = 3^{3^3} = 3^{27}$.
$9^{(2)} = 9^9 = (3^2)^9 = 3^{18} < 3^{27} = 3^{(3)}$.

$$3^{(4)} = 3^{3^{3^3}} = 3^{3^{27}}.$$
$$9^{(3)} = 9^{9^9} = (3^2)^{9^9} = 3^{2 \times 9^9}.$$

$3^{27} > 9^9$ from earlier but is $3^{27} > 2 \times 9^9$?

UP3
We would like to use the heuristic *aiming for sub-goals* and conjecture that $3^{27} > 2 \times 9^9$.

DP3
We want to find out without using the calculator why $3^{27} > 2 \times 9^9$, by manipulating the expressions to obtain the same base.

CP3
$2 \times 9^9 = 2 \times (3^2)^9 = 2 \times 3^{18}.$
Clearly, $3^{27} > 3^{19} = 3 \times 3^{18} > 2 \times 3^{18}.$

UP4
We are almost ready to solve the original problem and conjecture that $3^{(n+1)} > 9^{(n)}$.

DP4
We want to find out without using the calculator why $3^{(n+1)} > 9^{(n)}$, by manipulating the expressions to obtain the same base.

CP4
$$9^{(n)} = (3^2)^{9^{(n-1)}} = 3^{2 \times 9^{(n-1)}}.$$

$3^{(n+1)} - 9^{(n)} = 3^{3^{(n)}} - 3^{2 \times 9^{(n-1)}} > 0$ if $3^{(n)} > 2 \times 9^{(n-1)}$.

So we see here there is some form of induction because the n-th step requires the result from the $(n-1)$-th step.

UP5
We conjecture that $3^{(n)} > 2 \times 9^{(n-1)}$.

DP5
We use MI.

CP5
For positive integers n, let $P(n)$ be the statement: $3^{(n)} > 2 \times 9^{(n-1)}$.
When $n = 1$, LHS $= 3 > 2 \times 9^0 =$ RHS. Thus, $P(1)$ is true.
Suppose $P(k)$ is true for some $k \geq 1$, i.e. $3^{(k)} > 2 \times 9^{(k-1)}$.
Now, we have $3^{(k+1)} - 9^{(k)} = 3^{3^{(k)}} - 3^{2 \times 9^{(k-1)}} > 0$ since $3^{(k)} > 2 \times 9^{(k-1)}$ (by induction hypothesis).
Thus, $3^{(k+1)} > 9^{(k)}$.
Now $9^{(k)} = (3^2)^{9^{(k-1)}} = 3^m$ for some integer m.
Hence we have $3^{(k+1)} > 9^{(k)} \Rightarrow 3^{(k+1)} > 3^m \Rightarrow 3^{(k+1)} \geq 3^{m+1} \Rightarrow 3^{(k+1)} \geq 3 \times 3^m > 2 \times 3^m = 2 \times 9^{(k)}$.
Since $P(1)$ is true, and $P(k)$ implies $P(k + 1)$, by MI, $P(n)$ is true for all $n \geq 1$.

UP6
We finally prove the original problem rigorously.

DP6
We adapt the result of *CP5* and include the trivial argument that $3^{(n)} < 9^{(n)}$.

CP6
For positive integers n, let $P(n)$ be the statement: $3^{(n)} > 2 \times 9^{(n-1)}$.
When $n = 1$, LHS $= 3 > 2 \times 9^0 =$ RHS. Thus, $P(1)$ is true.

Suppose $P(k)$ is true for some $k \geq 1$, i.e. $3^{(k)} > 2 \times 9^{(k-1)}$.

Now, we have $3^{(k+1)} - 9^{(k)} = 3^{3^{(k)}} - 3^{2 \times 9^{(k-1)}} > 0$ since $3^{(n)} > 2 \times 9^{(n-1)}$ (by induction hypothesis).

Thus, $3^{(k+1)} > 9^{(k)}$.
Now $9^{(k)} = (3^2)^{9^{(k-1)}} = 3^m$ for some integer m.

Hence we have $3^{(k+1)} > 9^{(k)} \Rightarrow 3^{(k+1)} > 3^m \Rightarrow 3^{(k+1)} \geq 3^{m+1} \Rightarrow 3^{(k+1)} \geq 3 \times 3^m > 2 \times 3^m = 2 \times 9^{(k)}$.

Since $P(1)$ is true, and $P(k)$ implies $P(k+1)$, by MI, $P(n)$ is true for all $n \geq 1$.

$3^{(n)} > 2 \times 9^{(n-1)} \Rightarrow 3^{(n)} > 9^{(n-1)}$. Since trivially, $3^{(n)} < 9^{(n)}$, we have that the least m is $n+1$.

The participants appreciated the heuristic *aiming for sub-goals* as it was in line with the emphasis on the induction step while using the MI technique. By putting it in the framework of Pólya's problem solving model, students may be led naturally to rephrase the problem to concentrate on that sub-goal.

4 Feedback and Summary

About two months after the workshop, feedback on the workshop was sought from the participants via email. Eleven responded, out of which one declined because he was not teaching in a junior college. Since only a small proportion of the 36 teachers replied, the results of the feedback should be treated with suitable caution. The 5 questions required Likert scale responses ranging from 1 (strongly disagree) to 5 (strongly agree). Table 2 below shows the questions and the mean responses.

Table 2
Feedback from workshop participants

Statement	Mean Response
There is enough time to effectively teach MI in the H2 syllabus.	3.1
We will need more time if we were to teach MI "by experiencing authentic problem solving" as discussed in the workshop.	4.2
In general, my students currently have no problem understanding MI.	3.5
I think that teaching MI "by experiencing authentic problem solving" is worth trying.	4.5
I personally find the workshop *"Teaching MI by experiencing authentic problem solving"* interesting.	4.6

Overall, the participants who responded enjoyed the workshop and thought that the method of teaching MI by experiencing authentic problem solving was worth trying. They agreed however that more time would be needed for such an approach. Within the current curriculum, they barely had time to cover this topic. Also noted is the 'average' understanding of the students regarding MI.

This chapter sets out to suggest a method for teaching MI in the schools. It agrees with Harel (2002) that the common method of trying to fix students' misconceptions may not be efficient as MI is a proof technique that requires students to experience mathematical proving. We go one step further and bring to the foreground problem solving which is supposed to take centre stage in the Singapore mathematics syllabus. By solving a problem, the student will learn when to 'naturally' use MI, i.e. when a conjecture is made on the basis of process pattern generalization. Also, the student will learn the useful heuristic of 'aiming for sub-goals' to concentrate his resources on the inductive step of the proof. This skill of using the heuristic may then be transferable to solving problems in other topics in mathematics. Finally, the pedagogy requires student participation in all aspects of the problem solving endeavour thus making it an impactful learning experience for him or her. The generally positive feedback from the participants indicate that this method of teaching MI is worthy of a trial.

Acknowledgement

The authors thank the workshop participants for their responses and feedback.

References

Avital, S., & Libeskind, S. (1978). Mathematical induction in the classroom: Didactical and mathematical issues. *Educational Studies in Mathematics, 9*, 429-438.

Baker, J. (1996). Students' difficulties with proof by mathematical induction. Paper presented at the *Annual Meeting of the American Educational Research Association:* New York.

Borasi, R. (1985). Using errors as springboards for the learning of mathematics: An introduction. *Focus on Learning Problems in Mathematics, 7(3 & 4)*, 1-14.

Brumfiel, C. (1974). A note on mathematical induction. *Mathematics Teacher, 67(7)*, 616-618.

Cai, J., & Brook, M. (2006). Looking back in problem solving. *Mathematics Teaching incorporating Micromath, 196 (May)*, 42-45.

Chow, M. K. (2002). *Mastery of mathematical induction among JC students.* Unpublished Masters dissertation: Nanyang Technological University.

Chow, M. K. (2003). Mastery of mathematical induction among JC students. *The Mathematics Educator, 7(2)*, 37-54.

Dubinsky, E. (1986). Teaching mathematical induction I. *Journal of Mathematical Behavior, 5(3)*, 305-317.

Dubinsky, E. (1989). Teaching mathematical induction II. *Journal of Mathematical Behavior, 8(3)*, 285-304.

Foret, K. F. S. (1998). *Teaching induction: Historical perspective and current views.* Unpublished PhD dissertation: The American University.

Harel, G. (2002). The development of mathematical induction as a proof scheme: A model for DNR-based instruction. In S. R. Campbell & R. Zazkis (Eds.), *Learning and teaching number theory* (pp. 185-212). Ablex Publishing: Westport, CT.

Kroll, D. L., & Miller, T. (1993). Insights from research on mathematical problem solving in the middle grades. In D. T. Owens (Ed.), *Research ideas for the classroom: Middle grades mathematics* (pp. 58-77). New York: Macmillan.

Krulik, S., & Rudnik, J. A. (1980). *Problem solving: A handbook for teachers.* Boston: Allyn & Bacon.

Lester, F. K. (1994). Musing about mathematical problem-solving research: 1970-1994. *Journal of Research in Mathematics Education, 25*, 660-676.

Longman Dictionary of Contemporary English (1978). Longman: Bath, UK.

Ministry of Education (2012). *O and N(A) level Mathematics teaching and learning syllabus.* Singapore: Author.

National Council of Teachers of Mathematics (1989). *Curriculum and evaluation standards for school mathematics.* Reston, VA: Author.

Toh, T. L., Quek, K. S., Leong, Y. H., Dindyal, J., & Tay, E. G. (2011). *Making mathematics practical: An approach to problem solving.* Singapore: World Scientific.

University of Cambridge Local Examinations Syndicate (1998). *Mathematics examination syllabuses* (for candidates in Singapore only). Cambridge: Author.

University of Cambridge Local Examinations Syndicate (2006). *Mathematics examination syllabuses* (for candidates in Singapore only). Cambridge: Author.

University of Cambridge Local Examinations Syndicate (2013). *Mathematics examination syllabuses* (for candidates in Singapore only). Cambridge: Author.

Chapter 12

Scaffolding and Constructing New Problems for Teaching Mathematical Proofs in the A-Levels

ZHAO Dongsheng

Though most people agree that training in proof and reasoning is necessary and important in school mathematics, the teaching of mathematical proofs is still not given sufficient emphasis in the A-Level curriculum. The teaching of mathematical proofs lacks a reasonable systematic implementation. In this chapter, we discuss two main issues related to the teaching of proofs in A-Level Mathematics: (1) how appropriate scaffolding can be crafted to guide students in learning mathematical proofs via mathematical problem solving; and (2) how proof problems can be constructed for use in teaching A-Level Mathematics. This chapter appends a list of proof questions which can be employed by teachers for instructional use as demonstrated in this chapter.

1 Introduction: Why Mathematical Proof?

One of the unique features of mathematical science is its heavy emphasis on logical reasoning. A claim is valid or accepted in mathematics, only if it can be proved through deductive reasoning. Logical deductive reasoning is a powerful tool for mathematicians to expand the domain of mathematical knowledge and discover new results. The primary means of helping students develop deductive reasoning and mathematical understanding is through the teaching of proof in the classroom (Hanna, 2000; Zaslavsky et al., 2012).

2 What are the Possible Challenges and Difficulties in Teaching Mathematical Proofs?

A quick study of the 2013 Advanced Level mathematics syllabus of the University of Cambridge Local Examinations Syndicate (2013) shows that mathematical proof, with the exception of mathematical induction, is not much emphasized. This is in spite of the fact that Mathematical Reasoning is one of the areas of emphasis in the recent curriculum review. In view of the importance of mathematical proofs in relation to the training of students in logical reasoning, teachers should be preparing their students in mathematical proofs. What exactly are the factors that are holding them back?

Firstly, engaging students in proving is comparatively more time-consuming compared to solving a close-ended mathematical problem. From a teacher's perspective, the time invested in such activities, which is not emphasized in the syllabus, can be better channelled to preparing students for examination-type questions. Secondly, the solution of a problem involving mathematical proof is not unique. Hence more time is needed to grade students' solution. Furthermore, it is not very easy to keep the assessment consistent and completely fair. Thus, assessment of students becomes an issue which teachers need to consider.

From the students' perspectives, there are two major difficulties involving mathematical proofs. Firstly, students often find it difficult to organize their arguments in a clear and logical manner, and to present their proofs coherently. They are often uncertain about how much details should be included in their proofs. Secondly, and most importantly, every problem involving mathematical proof is different from every other problem. There is no routine method to follow for solving such problems. As such, students require more time and more persistence in order to find a solution.

In fact, from the author's perspective, it is precisely because of this last factor that teaching mathematical proofs is valuable even at the A-

Levels: it is an opportunity for students to see the relevance of learning mathematical problem solving, which is the heart of the Singapore mathematics curriculum. In other words, the bridge between mathematical proofs and the school mathematics is Mathematical problem solving.

3 Problem Solving and Scaffolding

3.1 *Problem Solving*

As we believe that it is crucial to link mathematical proof to problem solving, we begin with a review on mathematical problem solving.

Any student attempting mathematical problem solving requires a model to which he or she can refer, especially when progress is not satisfactory. Good problem solvers would presumably have built up their own models of problem solving.

A problem solving model that is made explicit to students should be helpful in guiding them in the learning of problem solving, and in regulating their problem solving attempts. Even a good problem solver may find the structured approach of a model useful. As Schoenfeld (1985) recounts in the preface to his book *Mathematical Problem Solving* about Polya's book *How to Solve It*:

> In the fall of 1974 I ran across George Polya's little volume, How to Solve It. I was a practising mathematician ... My first reaction to the book was sheer pleasure. If, after all, I had discovered for myself the problem-solving strategies described by an eminent mathematician, then I must be an honest-to-goodness mathematician myself! After a while, however, the pleasure gave way to annoyance. These kinds of strategies had not been mentioned at any time during my academic career. Why wasn't I given the book when I was a freshman, to save me the trouble of discovering the strategies on my own? (p. xi)

The practical approach which we would describe later uses Polya's model as the problem solving model made explicit to students. Among other reasons for our choice, the model is well-known and it is mentioned in the syllabus document of the Singapore Ministry of Education (2012). We also wanted a model which is easy for students to "carry about" in their heads. The essential features of Polya's problem solving model is shown below (Figure 1). The model is depicted as a flowchart with four components, *Understand the Problem, Devise a Plan, Carry out the Plan*, and *Check and Extend*, with back-flow allowed to reflect the dynamic nature of problem solving.

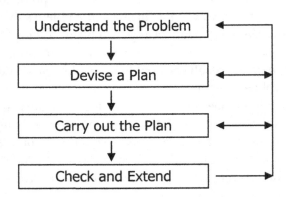

Figure 1. Flowchart of Polya's Problem Solving Model

3.2 Scaffolding

We believe that providing appropriate scaffolding is inevitable in teaching mathematical proofs. From the literature on mathematics education, it is clear that scaffolding is required when students encounter a situation they find difficult to understand or a problem that they are unable to solve by their own "unassisted efforts" (Wood, Bruner, & Ross, 1976).

Scaffolding is an analogy taken from the construction industry, in which a tool is used for workers to construct buildings. This tool consists of a structure placed against the building to support the workers to enable them to reach places that are otherwise inaccessible.

Using this analogy, scaffolding in education allows learners to reach places they would otherwise be unable to reach. Scaffolding could be in the form of hints, guides or questions. Wood, Bruner and Ross (1976) identify six key functions of scaffolding in education:

- Engage the learner in an interesting and meaningful activity;
- Develop the activity around manageable components;
- Ensure that the learner is on-task for a solution;
- Accentuate the main parts of the activity;
- Reduce the frustration level on the part of the learner; and
- Provide a model of the solution method for the learner.

Toh et al. (2011) discuss the role of scaffolding provided by the teachers when students are solving unfamiliar problems and got stuck. Teachers should understand that the purpose of scaffolding is to improve the students' learning processes rather than to provide overt help on the correct method of solving a given problem. Three levels of scaffolding were distinguished: (1) general – emphasis on the general processes; (2) specific – emphasis on the method to solve the same type of problems; and (3) problem-specific – what to consider in solving that particular problem. Teachers should begin with the general scaffolding when students first ask for help; then move on to subsequent levels, as necessary, if the students continue to have difficulties with the problem.

4 Mathematical Tasks derived from Problems Involving Proofs

In this section, we shall demonstrate how teachers can design the appropriate scaffolding to facilitate students to learn the processes of mathematical proof through a typical problem.

Scaffolding crafted from problems involving mathematical proofs provide students with rich learning experiences that can serve to nurture students' learning of mathematics in two ways: firstly, the tasks could be used to develop students' understanding of mathematical proof. Secondly, the learning experience acquired through these activities develops students' higher order thinking skills in mathematics. Not only that, students could even be challenged to re-examine their previously acquired understanding and belief in mathematics.

Problems in mathematical proof usually involve some known results in mathematics. Hence, the end result of the "problem" is already known. Thus, in engaging students to learn mathematical proof, we are not demanding students to "discover" new results; rather, we are equipping them with the thinking processes and learning experiences that mathematicians use in discovering new results in mathematics.

We see this as a good opportunity for students to be engaged in the entire problem solving process and the various heuristics in mathematical problem solving, such as "working backward" and "using an algebraic equation". We shall illustrate with two problems in mathematical proofs and how we can provide the appropriate scaffolding to facilitate students' learning.

Problem A-1.
Prove that if m and n are even integers, then $m+n$ is an even integer.

Problem A-2.
Prove that if m and n are odd integers, then mn is an odd integer.

Task A-1:
In this activity, we will prove that if m and n are even integers, then $m+n$ is an even integer.

Stage I: Do you understand the problem?
a)   Do you have difficulty understanding the statement of the problem? If so, how did you overcome it?

b)   Do you believe that the given statement is correct? Let's try some numbers to convince yourself that it is *possible* that the statement could be correct before we attempt to prove it.

c)   What is/are the given condition(s)? What is/are the result(s) that you are required to prove?

Stage II: Devise a plan
  a) State briefly how you think you can reach the result(s) from the given condition(s).

Stage III: Carry out the plan
  a) For each of the given conditions, use mathematical equations to represent them.

  b) Write down the equation representing the final result(s) that you are required to prove.

  c) Start from the given conditions in your equation and try to reach the final result. Show your working clearly for each step.

Step IV: Check and Expand
  a) Are you convinced that your proof is correct? Why?

  b) Try to create at least one other proof problem that is similar to the given problem. Try to prove it using the above procedure and check whether you can obtain the correct proof.

It is instructional to observe that the above scaffolding format can be used for Problem A-2. In fact, it is not difficult to see that the above scaffold via developing Polya's four stages can hold for many mathematical proofs that fall under the above category: for example,
  • The sum of two rational numbers is again rational;
  • The product of two rational numbers is again rational.

We leave the reader to adopt or modify the above scaffold for the following problems.

Problem B-1.
Prove that if $k \mid (m+2n)$ and $k \mid (2m+3n)$, then $k \mid m$ and $k \mid n$.

Problem B-2.
Show that for any integer n, $n^3 - n$ is divisible by 3.

Problem B-3.

Show that for any integer n, $n^5 - n$ is divisible by 30.

We shall demonstrate the scaffolding for another problem.

Problem C-1.

Suppose that a, b and c are positive integers such that $a \mid b$. Prove that $a \mid (b+c)$ if and only if $a \mid c$.

(Note: $a \mid b$ denotes the statement "b is divisible by a".)

Task C-1:
In this activity, we are given that a, b and c are positive integers such that $a \mid b$. Then we will prove that $a \mid (b + c)$ if and only if $a \mid c$.

Stage I: Do you understand the problem?

  a) Do you have difficulty understanding the statement of the problem? If so, how did you overcome it?

  b) Do you believe that the given statement is correct? Let's try some numbers to convince yourself that it is *possible* that the statement could be correct before we attempt to prove it.

  c) What is/are the given condition(s) for the whole problem? What are the two mathematical results that you have to prove? Paraphrase the statement into the two mathematical statements indicated by "if and only if"

Stage II: Devise a plan

  a) State briefly how you think you can reach the result(s) from the given condition(s).

Stage III: Carry out the plan

  a) For each of the given conditions, use mathematical equations to represent them.

  b) Beginning with the statement that you think is easier to prove, write down the equation representing the final result(s) that you are required to prove.

c) Start from the given conditions in your equation and try to reach the final result. Show your working clearly for each step.

d) Repeat for the second statement.

Step IV: Check and Expand
a) Are you convinced that your proof is correct? Why?

b) Try to create at least one other proof problem that is similar to the given problem. Try to prove it using the above procedure and check whether you can obtain the correct proof.

Observe that the problem C-1 is slightly more advanced compared to the previous task and the problems B-1 to B-3, as it involves both necessary and sufficient conditions. Thus, it is useful to lead students to *paraphrase* this compound statement into two mathematical statements of the "if-then" form.

5 Some Topics for Mathematical Proofs

In this section, we consider some topics, where we can provide the above scaffolding for students using the problem solving framework. The resulting activities are meaningful for students and should facilitate the learning of mathematical proofs required in the A-Level curriculum.

5.1 *Classification of integers*

The most basic classification of integers groups integers into two parts: even and odd numbers. Based on this classification, we can formulate many simple proof questions that reinforce the problem solving heuristic of "classification" emphasized in the school mathematics curriculum. Examples of problems are shown as follows.

Problem 5.1-1.
Prove that for any integer m, $m^3 + m^2 + m$ is odd if and only if m is odd.

Problem 5.1-2.
Prove that for any integer m and positive integer n, the number m^n is even if and only if m is even.

5.2 Divisibility and using algebraic representation

This is another rich topic for including simple proof questions. Many of these proofs require the representation of the information using algebraic representation such as equations, and algebraic manipulation, such as factorization. An example of such a problem is shown below.

Problem 5.2-1.
Prove that for any integer n, $2 \mid n$ if and only if $8 \mid n^3$.

Problem 5.2-2.
Prove that for if the integer n is composite, then $2^n - 1$ is also composite.

5.3 Rational and irrational numbers and other methods of proofs

While the concept of rational numbers is taught in secondary mathematics, it is seldom emphasized as there are relatively few questions in the high-stake national examinations. It is thus a good opportunity to reinforce these ideas at the A-Level through mathematical proofs via mathematical problem solving.

In addition, it could also be opportune to introduce the technique of proof by contradiction, adding to the students' repertoire of mathematical proofs (mainly, mathematical induction). A well-known example is the classic proof of the irrationality of $\sqrt{2}$. The following are some more examples of problems amenable to proof by contradiction.

Problem 5.3-1.
Let a be a rational number. Prove that for any real number b, $a+b$ is rational if and only if b is rational.

Problem 5.3-2.
Prove that for any non-zero rational number x and irrational number y, the number xy is irrational.

Problem 5.3-3.
It is known that $\sqrt{6}$ is irrational. Prove that $\sqrt{2}+\sqrt{3}$ is irrational.

5.4 Reinforcing mathematical concepts taught in the A-Levels

Mathematical proofs at the A-Levels can also tap on and reinforce the concepts taught in the A-Level mathematics syllabus. For example, the absolute value function and inequalities.

Problem 5.4-1.
Prove that for any numbers x and y, if $|x|/(|x|+1)=|y|/(|y|+1)$ then $x = y$.

We leave it to the reader to adapt the scaffolding discussed in the preceding sections to facilitate their students' learning. We would like to highlight that Problem 5.4-1 is an example of a proof question involving two topics, namely, (1) absolute value functions and (2) functions and graphs. Teachers could even guide the students to appreciate that Problem 5.4-1 can be paraphrased as:

Prove that the function $f(x)=|x|/(|x|+1)$ is a one-to-one function.

Problem 5.4-2.
Prove for any numbers x and y, that $\min\{x, y\}= \dfrac{x+y}{2} - \dfrac{|x-y|}{2}$.

Note that Problem 5.4-2 is a useful problem in engaging the students to apply the heuristics of "substituting suitable values" to understand a "new" function, thereby using problem solving strategies to learn new mathematics.

We append few more questions for the readers to analyze the underlying intent of these problems and to modify and adapt the scaffold discussed in the preceding sections for teaching mathematics in the A-Level classrooms.

Problem 5.4-3.
Show that for any positive real numbers a, b and c,
$$\frac{a+b+c}{3} \geq \sqrt[3]{abc}.$$

Problem 5.4-4.
Prove if $|x| < 1$, $|y| < 1$ and $|z| < 1$, then $xy + yz + zx > -1$.

Problem 5.4-5.
Show that for any real value x, $\sin^8 x + \cos^8 x \geq 1/8$.

Problem 5.4-6.
Prove that for any integer $n > 1$,
$$\log_n (n+1) > \log_{n+1} (n+2).$$

Problem 5.4-7.
Prove that if $a > b > c$, then
$$\frac{1}{a-b} + \frac{1}{b-c} + \frac{1}{c-a} > 0.$$

Problem 5.4-8.
Prove that if x, y and z are positive numbers, then
$$\frac{z}{x+y} + \frac{y}{x+z} + \frac{x}{y+z} > \frac{3}{2}.$$

6 How to Find and Create New Proof Questions?

We have discussed how mathematical proofs can be incorporated into the A-Level mathematics syllabus by finding its relevance in the problem solving curriculum and through using appropriate scaffolding. The next question to discuss is how one can find or create various types of proof questions with different levels of difficulties, both for teaching and assessment. In the above scaffolding, we have seen that Stage IV is an attempt to guide the students to "generalize" or create their own "new" problems. For students, we want to challenge them to not stop at a particular proof but to go beyond and pose a new problem. Teachers, on the other hand, must be able to create meaningful and pedagogically sound questions. However, we must admit that creating good and novel proof questions is difficult; much practice and experience is required.

In this section we shall propose some strategies for creating new questions, illustrated by concrete examples. We hope these will inspire more ideas from the readers.

6.1 *Generalization*

Generalization is one of the main motivations that lead to the discovery of new theories and results in mathematics. Based on a given question, we can try to extend or generalize it to obtain new one.

For instance, consider the following problem:

Problem 6.1-1.

Prove that if $x \neq y$, then $\dfrac{x}{|x|+1} \neq \dfrac{y}{|y|+1}$.

By changing the number 1 to an arbitrary positive number a or introducing another variable b, we obtain the following generalizations.

Problem 6.1-2.

Prove that if $a > 0$ and $x \neq y$, then $\dfrac{x}{|x|+a} \neq \dfrac{y}{|y|+a}$.

Problem 6.1-3.

Prove that if $a > 0$, $b \neq 0$ and $x \neq y$, then $\dfrac{x}{|bx| + a} \neq \dfrac{y}{|by| + a}$.

From the proof of "$\sqrt{2}$ is irrational", we naturally deduce the following general problem and its generalization to multiple primes.

Problem 6.1-4.

Let p be an arbitrary prime number. Prove that $\sqrt{p}$ is irrational.

Problem 6.1-5.

For any distinct primes p_1, p_2, ..., p_n, prove that $\sqrt{p_1 p_2 \cdots p_n}$ is irrational.

We illustrate with one more example where we generalize a problem involving two variables to one involving n variables.

Problem 6.1-6.

If $a, b \geq 0$ and $a + b = 1$, prove that $1 \leq \sqrt{a} + \sqrt{b} \leq \sqrt{2}$.

Problem 6.1-7.

If a_1, a_2, ..., $a_n \geq 0$ and $a_1 + a_2 + ... + a_n = 1$, prove that

$$1 \leq \sqrt{a_1} + \sqrt{a_2} + \cdots + \sqrt{a_n} \leq \sqrt{n}.$$

6.2 Specialization

Sometimes, we can also formulate some specific results from a general one. Students can be asked to prove these specific problems either directly, or to deduce them from the general one, if they recognize the connection to the general result. For instance, based on the Cauchy-Schwarz inequality, one can formulate many different specific inequalities.

Problem 6.2-1.

Prove that for any positive numbers a, b, c, $\dfrac{9}{a+b+c} \le \dfrac{1}{a} + \dfrac{1}{b} + \dfrac{1}{c}$.

Problem 6.2-2.

Prove that for any positive numbers a, b, c, and $l = \dfrac{a+b+c}{3}$, we have

$$\frac{a+b+c}{2} \le \frac{a^2}{a+l} + \frac{b^2}{b+l} + \frac{c^2}{c+l}.$$

Problem 6.2-3.

For any positive integer n and $0 \le x \le 1$, $\left(1 + \dfrac{x}{n}\right)^n \le \left(1 + \dfrac{x}{n+1}\right)^{n+1}$.

6.3 Modification

A third strategy is to modify an existing problem by replacing some parts of the problem with an equivalent condition. For example, consider the following problem.

Problem 6.3-1.

Prove that if $\alpha + \beta = \pi/2$, then $\sin^2 \alpha + \sin^2 \beta = 1$.

The condition "$\alpha + \beta = \pi/2$" is equivalent to "$\sin(\alpha + \beta) = 1$, $\sin \alpha \ge 0$ and $\sin \beta \ge 0$." Using the addition formula, we obtain "$\sin \alpha \cos \beta + \cos \alpha \sin \beta = 1$". If we now set $a = \sin \alpha$ and $b = \sin \beta$, we obtain the following problem.

Problem 6.3-2.

Prove that if $a\sqrt{1-b^2} + b\sqrt{1-a^2} = 1$, then $a^2 + b^2 = 1$

Finally, we recommend several journals that contain problems in elementary mathematics that may be suitable for A-Level students. These are "*The Mathematical Gazette*", "*American Mathematics Monthly*", "*The College Mathematics Journal*", and "*Mathematical Medley*".

7 Conclusion

In this chapter, we discussed how mathematical proof can be made relevant in the A-Level mathematics curriculum through providing appropriate scaffolding. The scaffolding was done by tapping on the problem solving framework of our curriculum. It should again be stressed that the whole objective of providing scaffolding is to empower the students to learn (Holton & Clarke, 2006), so that ultimately they can handle mathematical proofs on their own. We also discuss strategies in how "new" meaningful mathematical proof questions can be constructed. We strongly urge the reader to attempt to design the appropriate scaffolding for the questions posed in the Appendix and to design "new" proof questions based on the above discussion.

References

Hanna, G. (2000). Proof, explanation and exploration: An overview. *Educational Studies in Mathematics*, 44(1-2), 5-23.

Holton, D., & Clarke, D. (2006). Scaffolding and metacognition. *International Journal of Mathematical Education in Science and Technology*, 37:2, 127-143.

Ministry of Education (2012). *O- and N(A) level mathematics teaching and learning syllabus*. Singapore: Author.

Schoenfeld, A. H. (1985). *Mathematical problem solving*. Orlando, FL: Academic Press.

Toh, T. L., Quek, K. S. Leong, Y. H., Dindyal, J., & Tay, E. G. (2011). *Making mathematics practical: An approach to problem solving*, Singapore: World Scientific Publishing.

University of Cambridge Local Examinations Syndicate (2013). *Mathematics examination syllabuses* (for candidates in Singapore only). Cambridge: Author.

Wood, D., Bruner, J. S., & Ross, G. (1976). The role of tutoring in problem solving. *Journal of Child Psychology and Psychiatry*, 17(2), 89-100.

Zaslavsky, O., Nickerson, S. D., Stylianides, A. J., Kidron, I., & Winicki-Lamndman, G. (2012). The need for proof and proving: Mathematical and pedagogical perspectives. In Hanna, G. & de Villiers, M. (Eds.), *Proof and Proving in Mathematics Education, New ICMI Study Series 15*. Netherlands: Springer.

Appendix

A. General functions

Problem A-1.
Show that for any function f, the function g defined as
$$g(x) = f(x) - f(-x)$$
is odd.

Problem A-2.
Prove that every function can be expressed as the addition of an even function and an odd function.

Problem A-3.
Use definition to show that the function $f(x)=x^2 - 2x - 3$ is decreasing on the interval $(-\infty, -1]$.

Problem A-4.
Use definition to show that the function $g(x)=x+1/x$ is increasing on the interval $[1, +\infty)$.

Problem A-5.
Prove that the function $f: R \rightarrow R$ is surjective, where $f(x) = 2+5x$.

Problem A-6.
Show that the function $h(x) = 1/x$ is a convex function on $(0, +\infty)$.

Problem A-7.
Prove that if f and g are increasing functions then $f + g$ is increasing.
Is it true that if $f + g$ is increasing then at least one of f and g is increasing?

B Trigonometry functions

Problem B-1.
Prove for any x, $\cos^4 x = \sin^4 x - 2\sin^2 x = 1$.

Problem B-2.
Prove that $\pi/2$ is the minimal positive period of the function
$$f(x) = |\sin x| + |\cos x|.$$

Problem B-3.
Let A, B and C be the three angles of a triangle. Prove that
$$\sin\frac{A}{2}\sin\frac{B}{2}\sin\frac{C}{2} \le \frac{1}{8}.$$

Problem B-4.
For any positive integer n and number a $(a \ne 2k\pi)$,
$$\sin a + \sin 2a + \cdots + \sin na = \frac{\sin\dfrac{n+1}{2}a}{\sin\dfrac{a}{2}} \cdot \sin\frac{n}{2}a.$$

Problem B-5.
Prove that
$$\frac{\tan x \, \sin x}{\tan x - \sin x} = \frac{\tan x + \sin x}{\tan x \, \sin x}.$$

Problem B-6.
Prove $3\sin x - \sin 3x = 2\sin x\,(1 + \cos 2x)$.

Problem B-7.
Prove $\sin\alpha\,(\sin\alpha + \sin 3\alpha + \sin 5\alpha) = \sin^2 3\alpha$.

Problem B-8.
Prove
$$\frac{1}{2} \le \sin^4\alpha + \cos^4\alpha \le 1.$$

Problem B-9.
Prove $2 \le (2 - \sin\theta - \cos\theta)(2 + \sin\theta + \cos\theta)$.

Chapter 13

Learning Number in the Primary School through ICT

Barry KISSANE

Information and communications technology (ICT), increasingly available at both home and school, offers new kinds of learning experiences for primary children. This chapter explores the distinctive contribution of ICT to learning experiences of students, focusing on those that are not readily available through traditional media of pencil and paper and organised practice of skills. According to a Becta report in the UK, arguments and evidence for using ICT in the number curriculum focus on the role of feedback, observation of patterns and relationships and the development of visual imagery. The chapter highlights examples that use virtual manipulatives on the Internet, other web-based experiences, and hand-held devices such as Apple's iPad® and iPod Touch® and modern scientific calculators. The focus of the examples is on the development and understanding of important ideas in the number curriculum, such as those related to number patterns, place value, decimals, fractions, factors and ratios, rather than to the refinement of arithmetical skills. New learning experiences using ICT demand a new kind of attention to the role of the teacher, in a variety of contexts, including individual ICT use, small-group use, whole-class use (e.g. through interactive whiteboards) and home use.

1 Introduction

The chapter is premised on a view that the most important role of ICT in mathematics is to help create *meaning*. Although ICT can be used for other purposes (e.g., drill and practice), we will focus on how it might help pupils to understand mathematics better. So the focus is on concepts and thinking, rather than on practice or computation.

We also focus on giving opportunities that are not already available to pupils, such as those that appear in explanations in a textbook or practice exercises for procedures that have been taught. If ICT does not provide something different from what is otherwise available to pupils, it is questionable whether it is a wise use of resources, including time.

A very useful summary of the opportunities related to ICT appears in a British document produced by Becta (a UK government agency that has since been disbanded as a result of structural governmental changes) in association with the Association of Teachers of Mathematics and the Mathematical Association (2009). This document identified and exemplified five major opportunities for children to benefit from the use of ICT in mathematics, describing them as 'entitlements'. Three of these are of particular importance to the learning of Number, and are elaborated a little below.

Learning from feedback
We have long understood that feedback is helpful for learning. ICT feedback is usually immediate and, because it comes from a machine and not a person, is both non-judgmental and impartial. This creates an opportunity for children to engage with ideas, contexts and questions, without the risks associated with being wrong in front of the teacher or others. If children are encouraged to develop expectations about their responses, they can use ICT as an opportunity to test and adjust those expectations, and develop a better understanding of something. This is not unlike the well-known opportunities that children regularly take advantage of in learning computer games.

Observing patterns and seeing connections

Patterns are an intrinsic part of mathematics at all levels. An ICT tool can provide a rich opportunity for students to generate and observe patterns in order to see the underlying mathematical relationships and connections between mathematical ideas. Once patterns are observed, children can use them to make conjectures and generalisations and test these. While activity of this kind is available without ICT, the capacity of ICT to respond quickly and reliably provides a different experience than is otherwise available.

Developing visual imagery

Especially in recent years, ICT devices such as computers and tablets have provided new visual opportunities for learners, so that learning mathematics is no longer restricted as much to symbolic and numerical representations. The fresh opportunities created by ICT go beyond providing visual images, however, as the images can often be generated and manipulated by children. The old adage that 'a picture is worth a thousand words' is pertinent here. ICT provides children with an opportunity to visualise mathematical ideas and relationships and develop a richer understanding than is possible with words and symbols alone.

The remaining two entitlements described by Becta (2009) are concerned with exploring data and with 'teaching the computer'. Although these are also potentially relevant to the number curriculum in particular, they are less central than the earlier three entitlements and thus are not treated here in detail. Although it is set in a UK context, the Becta (2009) document provides many examples of these opportunities that Singapore teachers will find informative and helpful.

2 ICT in the Singapore School Mathematics Curriculum

The Primary Syllabus (Ministry of Education, 2012) makes it clear that ICT is key to the learning of mathematics. This is clearly stated in the Foreword: Learning Mathematics – A 21st century necessity:

… The curriculum must engage the 21st century learners who are digital natives comfortable with the use of technologies and who work and think differently. The learning of mathematics must take into cognizance the new generation of learners, the innovations in pedagogies as well as the affordances of technologies. (p. 3)

The Mathematics Framework does not refer extensively to the use of ICT by pupils or teachers, but it is clear that there is an implied role in several components. For example, mathematical concepts are the building blocks upon which mathematical procedures and skills, applications of mathematics and problem solving in mathematics all depend. In describing the development of concepts, the Syllabus specifies a role for ICT:

> To develop a deep understanding of mathematical concepts, and to make sense of various mathematical ideas as well as their connections and applications, students should be exposed to a variety of learning experiences including hands-on activities and the use of technological aids to help them relate abstract mathematical concepts with concrete experiences. (p. 17)

In this context, it seems appropriate to extend the meaning of 'hands-on activities' to include not only working with physical materials but also 'virtual' ones, which also provide concrete experiences for students, albeit in the virtual rather than the physical world.

Similarly, the Framework identifies a suite of key skills associated with mathematics and of importance to mathematical problem solving.

> In today's classroom, these skills also include the abilities to use spreadsheets and other software to learn and do mathematics. (p. 17)

A good example of this is the recent change to include calculator use in mathematics for Primary 5 and 6, in recognition that calculators and other technology tools are both tools for learning mathematical concepts as well as tools for deploying and enhancing mathematical skills. While calculators are widely recognised as helpful devices to undertake numerical calculations, it is important to balance calculation by machine with other ways of completing calculations. Thus, in describing mathematical modelling, the Syllabus recognises explicitly the need for students to "select and use appropriate mathematical methods and tools (including ICT)" (p. 16), implicitly suggesting both that students have a repertoire of methods at their disposal and are supported to make good choices amongst them in practice.

The role of calculators for learning mathematics, and not just for calculating, is less widely appreciated and so is described briefly later in this chapter.

The Primary Syllabus highlights the importance of ICT for teaching mathematics in "Principle 3: Teaching should connect learning to the real world, harness ICT tools and emphasise 21st century competencies". (p. 23). In elaborating this principle, the Syllabus observes:

> Teachers should consider the affordances of ICT to help students learn. ICT tools can help students understand mathematical concepts through visualisations, simulations and representations. They can also support exploration and experimentation and extend the range of problems accessible to students. The ability to use ICT tools is part of the 21st century competencies. (p. 24)

While the Syllabus is generally encouraging for the use of ICT for learning mathematics in Singapore primary schools in the 21st century, informal discussions with primary teachers suggest that encouragement alone may not be sufficient. Many teachers report feeling pressured to complete the Syllabus and to prepare children for important examinations, and there continues to be a focus on traditional paper and pencil work in mathematics in many classrooms. Other teachers report relatively little experience using ICT for learning purposes, so that one intention of this chapter is to offer advice on suitable sources of materials for the number aspects of the primary curriculum.

3 Contexts for ICT Use

Experience with ICT will not by itself bring about pupil learning, so the work of the teacher to focus attention appropriately and provide a context for learning needs to be recognised. There are at least three different contexts (described below as *personal*, *pairs* or *plenary*) in which learning with ICT might take place:

- *Personal*: Pupils might use a personal ICT device (such as a fixed computer in a lab, portable laptop computer, *iPad* or other tablet PC, scientific calculator) by themselves; some of this use might occur at home (e.g., as part of a homework task), as well as at school.
- *Pairs*: Pairs or small groups of pupils might use ICT together, so that they are engaged in an activity with their peers, involving discussion and mathematical thinking together. In some cases, this may involve joint use of an interactive whiteboard, while in other cases, a shared computer or calculator is appropriate.
- *Plenary*: Whole class ICT use, such as a class discussion might be stimulated by use of a projected image, an interactive whiteboard, or both.

Physical and educational contexts in different schools are different, and facilities for ICT vary both from school to school and from pupil to pupil. Similarly, confidence in and the use of ICT will vary from teacher to teacher, and even from year level to year level.

Informal discussions with Singapore primary teachers reveal that data projectors and computers in classrooms are in widespread use in classrooms, so that the necessary facilities for plenary ICT activities are available in many, if not most schools. In recent years, it seems that tablet PCs, such as *iPads*, have been adopted in some primary settings, increasing the opportunities for both personal and paired work with ICT for learning. By their nature and size, calculators lend themselves to personal or paired use, although projected versions of calculators via computers allow for plenary discussion as well. Many Singaporean teachers also report high levels of computer access in the homes of primary children, making possible new kinds of learning at home.

To some extent, inevitably, readers need to interpret this chapter in their own context. However, it seems that Singapore schools overall enjoy many opportunities for the use of ICT in learning mathematics that are not shared by some neighbouring countries, as both schools and individuals are relatively well resourced as far as ICT is concerned.

4 Sources of ICT on the Internet

Like other modern curricula, the Singapore Primary syllabus clearly intends for effective use of ICT for student learning, although the finer details of the role of digital technologies for learning are not evident in the official documents. (Ministry of Education, 2012). Recent and substantial movements in the access to the Internet in schools and homes have made realistic the prospects of using the Internet as a resource for both teaching and learning. (Kissane, 2009). This section explores briefly some of these new opportunities, with a particular focus on virtual manipulatives. In addition, the emphasis is on those available at minimal (or no) cost and likely to be helpful for pupil learning in Number in particular. Many suitable resources can be accessed via a website (Kissane, 2013a), although it needs to be recognised that there is always some instability around web offerings, and that sometimes resources are moved because the publishers have chosen to move them.

While some virtual manipulatives are provided by publishers, to accompany purchased textbooks or interactive whiteboards, the focus here is on those that are publicly available via an Internet connection and a recent web browser. Typically, computers need to have some software installed that allows the dynamic aspects to work, with the two most likely platforms being Java and Flash. Very often these days, modern computers are set up to load such materials so that the user is not even aware of what is happening (not unlike drivers being unaware of the subtleties of their gearbox in an automatic car). Most modern browsers and online sites will help users deal with the problems (such as locating, installing and activating suitable software plug-ins), so that extensive technical help is not always needed. Sometimes (as in the case of the *iPad*, for example), however, the problems are presently irresolvable.

There are many possible sources of virtual manipulatives, and this chapter certainly does not claim to exhaustively catalogue them. Kissane (2013b) describes a number of sites that have an interactive component, many of which contain virtual manipulatives; this site also links to another website maintained by the author, with both of these described a little in Kissane (2009). The Center for Implementing Technology in Education (2012) also provides a helpful description of some sources of virtual manipulatives.

Systematic and structured collections of virtual manipulatives provide many good examples for primary teachers. One of the earliest such collections, popular with many teachers, is the National Library of Virtual Manipulatives, constructed and maintained at Utah State University (2010). The collection of more than 100 Java applets spans a range of strands of mathematics (one of which concerns Numbers and Operations) across a range of year levels. While the classification into year levels is at times questionable, teachers will find good examples here to suit many purposes. Many applets include activities for students to undertake and all include good advice for students, teachers and parents to make the intentions and operations clear. An additional feature of this collection is that it can be purchased for storage onto a computer, so that it is accessible without Internet access, which can be very helpful when access is problematic, intermittent or slow. Good examples of applets appropriate for learning Number in the primary school are *Factor Tree*, *Fractions* (various), *Place Value Number Line*, *Base Blocks*.

Another older set of manipulatives called *Project Interactivate* was developed by the Shodor Educational Foundation (2012). It also comprises more than 100 separate manipulatives, with good supporting advice and lesson ideas for teachers. Interestingly, recent initiatives have been to develop some materials specifically for mobile platforms, such as the *iPod Touch* and *iPad*, perhaps a sign of things to come. Good examples of suitable materials related to Number are *Bounded fraction finder*, *Fraction sorter*, *Estimator*, *Mixtures*.

The National Council of Teachers of Mathematics (2012) has developed the highly-regarded *Illuminations* site over recent years to support its other national curriculum initiatives. This collection is also well-structured in the sense that it is organised into grades and content areas and can be searched. Applets include adequate advice on how to use them as well as describing their educational significance. Good examples from this collection related to Number include *Equivalent fractions, Grouping and grazing, Factorize.*

The *Nrich* site from Cambridge University in the UK provides a wealth of materials of many different kinds for learners of mathematics across a wide range. This site is regularly updated and can be easily searched to locate materials for a particular mathematical topic. Many of the interactive materials require Flash capabilities (and hence will not work on some tablets, such as *iPads*). A few packages have been developed for teachers to download and use, so that Internet access is not required for classroom use; a good example of this is *Introduction to Fractions* (University of Cambridge, 2013)

In addition to freely available materials, some good materials are available only to subscribers (a timely reminder, if one is necessary, that the costs of producing quality learning materials online is high and requires resources to be made available somehow if it is to be accomplished effectively). A good Australian example is the *Maths300* collection of lesson materials, which also includes a set of innovative software that can be downloaded for subscribers to use. (Education Services Australia, 2010) The software is mostly intended for teacher use with a class, and much can be used very effectively on an interactive whiteboard. The lessons themselves are a wonderful collection, based on the work of Australian teachers; many of the lessons make it clear how the software can be effectively used in classrooms, and draw on real experience in Australian classrooms. A subscription to *Maths300* is a very worthwhile investment for any school.

Similarly, the *HOTmaths* online learning system (Cambridge University Press, 2012) includes many interactive objects in the form of virtual manipulatives (called widgets). Some of these have a role of explaining and demonstrating ideas, while others offer opportunities for genuine manipulation and engagement in nicely targeted ways. Although a subscription is necessary to access the materials, the web site allows visitors to engage with a sample of widgets to get a feel for the kinds of ideas involved.

4.1 Virtual manipulatives

For some time now, particularly in primary and middle schools, the important place of student engagement with manipulatives has been recognised. In the space of a generation or two, we have moved from mathematics as being regarded as an activity that is solely conducted with pencil and paper, and with seemingly endless expectations for drill and practice. Modern teachers and modern classrooms have routinely sought ways to engage students in doing something active to stimulate, support, enhance and encourage their learning of important concepts. Over a slightly shorter period, computers have emerged in schools and homes as significant elements of the learning infrastructure, so that today most classrooms and many homes have a visible ICT presence.

The idea (and indeed, even the term) of virtual manipulatives for mathematics is probably around a decade old now, according to a very good early paper describing them (Moyer, Bolyard & Spikell, 2002). Essentially, the paper highlighted that virtual manipulatives share many characteristics of concrete manipulatives, yet are available on a computer screen rather than in tangible form. Importantly, the digital objects include visual displays that can be manipulated by students in various ways, usually via pointing, clicking or dragging with a mouse, which is what gives them similar potentials for learning to manipulations of physical objects (like blocks, straws, scales, rulers, etc …). Indeed, Moyer et al (2002, p. 373) suggest that the term 'virtual manipulative' not be used for objects that are not genuinely manipulable, such as those that engage students in merely answering questions posed by a computer program displaying static images, similar to those on a printed page.

This distinction suggests that not all digital objects can be classified as virtual manipulatives; some serve other roles, such as demonstration or presentation of mathematical material, albeit in an apparently dynamic way. The extraordinary *Nrich* website (University of Cambridge, 2012) similarly contains virtual manipulatives, together with other kinds of digital objects that are not intended for students to manipulate.

As the Center for Implementing Technology in Education (2012) observes in a nice recent summary of work with virtual manipulatives, there is relatively little research on the effectiveness of the idea yet, since it has been in use for only a few years; this chapter does not claim to summarise this research systematically. However, early signs are promising that, not only do virtual manipulatives enhance the classroom experience of children and their teachers, but they may make distinctive contributions to helping students understand important mathematical concepts and relationships, if well managed by teachers, and certainly are consistent with the Primary syllabus aspirations for the use of ICT.

4.2 Roles for manipulatives

Many structured collections of virtual manipulatives include detailed advice for teachers, highlighting some of the ways in which the manipulatives might be used. Some even include particular activities or suggested tasks for students to undertake, while others leave most such details to creative teachers and curious students. In general, it seems ambitious to expect students to learn mathematics from virtual manipulatives without advice of some kind regarding what to do, however, so that some care is needed to ensure that the mathematical ideas are extracted highlighted appropriately from any activity. It is also remarkably easy to forget that the experience of using a virtual manipulative by someone already familiar with the mathematical idea involved (notably, the teacher) is not the same as the experience of someone new to the idea (notably, the student), so that care is needed to make sure that students are thinking about what they are doing in a way that is likely to be helpful to them. This is not a new idea: Wheatley (1992) expressed similar concerns regarding manipulatives in general, highlighting the importance of students reflecting on their activity.

As an illustration, Figure 1 shows the virtual manipulative *Base Blocks* from the National Library of Virtual Manipulatives.

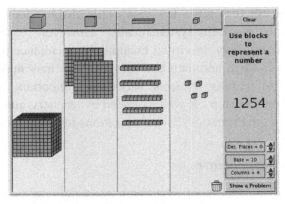

Figure 1. Representing numbers with *Base Blocks* (Utah State University, 2010)

This manipulative allows young children to represent numbers with (virtual) objects, highlighting the nature of place value representations in a highly visual way. While teachers already familiar with place value might not at first recognise the significance of such a representation, it is potentially very powerful for young children. While blocks of these kinds have been used previously to help understand place value, the virtual manipulative in this case provides opportunities not normally available: as blocks are moved to the mat, the numbers are routinely shown, which helps students to appreciate the critical patterns involved. Additionally, if a block is moved to the adjacent column on the right, it is routinely broken down, making clear the patterns of tens involved; similarly, collections of ten objects can be grouped together and moved to the adjacent column on the left. The teacher needs to plan carefully to highlight the mathematics they want students to learn; in this case, the manipulative can be also used for bases other than ten, or to represent decimal numbers as well as whole numbers, allowing students to understand deeply the place value concepts involved. Young children can be helped by reducing the number of columns on display. Several versions of this manipulative are provided on the web site, to allow it to support addition and subtraction as well as number representations.

A key role of manipulatives, virtual or otherwise, is to give students direct personal experiences of a concept or a relationship. If an object can be manipulated, then there is some potential for students to learn something by watching the effects of their manipulations, making and testing predictions of what will happen, discussing their observations with others or systematically recording the results of their manipulations in some way. So, some of the classroom potential for virtual manipulatives is similar to the potential of other activities involving an element of students discovering things for themselves. This offers a different and complementary kind of learning experience than being told about something by a textbook or by a teacher, which is still a common mode of learning for much of mathematics.

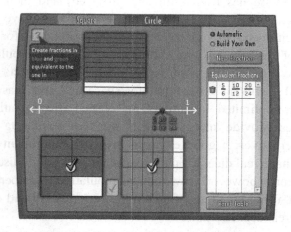

Figure 2. Exploring *Equivalent Fractions*
(National Council of Teachers of Mathematics, 2012)

Consider Figure 2, for example, which shows *Equivalent Fractions* from the *Illuminations* website (National Council of Teachers of Mathematics, 2012). This manipulative shows physically some key ideas: that the same number can be represented by a fraction in different ways, yet the number has only one location on the real number line. Children using the manipulative are required to construct for themselves two different fractions (with different denominators) to match a given fraction. Visual queues are provided with the associated number line. When the task is completed, results are routinely tabulated, so that the patterns associated with equivalent fractions can be scrutinised.

To focus attention on the important ideas here, teachers might suggest that children use the resulting table of values to see the patterns involved, to help them understand why 'cancelling' fractions is a helpful operation. A virtual manipulative of this kind might be used personally, in pairs or in plenary mode, perhaps using a public interactive white board. Using virtual manipulatives requires teachers to think about what the students will actually do, as for any other kinds of classroom activity.

Some virtual manipulatives are similar to concrete manipulatives, but offer clear advantages in terms of their cost (essentially free, after the ICT infrastructure has been built and provided) and their availability (essentially exhaustive). Thus, while many classrooms have a supply of multibase arithmetic blocks to share among students for use in an assortment of productive ways, a comparable virtual manipulative has an inexhaustible supply of these for each student. Other objects are often in more limited supply in classrooms, for space and cost reasons, yet are available in unlimited quantities in virtual manipulatives. Other manipulatives provide more efficient ways of exploring ideas than concrete alternatives; a good example involves experiments with dice. Classrooms have real dice, which students can and should use, to explore random events; however a virtual manipulative can permit similar explorations, using hundreds of dice rolls, very quickly and efficiently – and with a great deal less noise!

As a third example of this important species of mathematical activity, Figure 3 shows a 'zoomable' number line from *MathsIsFun*.

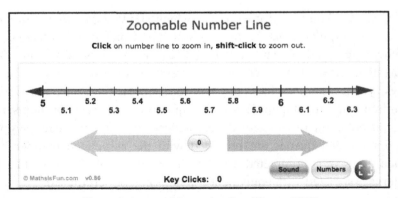

Figure 3. A zoomable number line (Pierce, 2013)

This number line shows integers and decimals, positive and negative, to essentially unlimited accuracy and can be scrolled to right and left and zoomed in and out. Children using this virtual manipulative can locate any real number, by scrolling and then zooming. They can also see that there are unlimited numbers between any two numbers (by zooming in at an appropriate place). The manipulative can be used successfully by individuals, by pairs of students or by a whole class of students. Consistent with the Becta document (2009) suggestions, it provides an interactive experience, a visual experience and also allows children to see the repeating patterns of the decimal number system. Clearly, it provides opportunities for learning that are not present with a static number line or with standard textbook exercises,

5 Interactive Whiteboards

Interactive whiteboards (IWB) have become popular in recent years, especially in primary schools, although they are not yet in widespread use in Singapore. Essentially, an interactive whiteboard is a touch-sensitive board large enough for the whole class to see that shows a projected computer screen, rather than using a mouse or keyboard. Manufacturers of particular boards provide educational software to suit the board, but it is also possible to use the board as a place to project whatever is on a computer, so that it is not necessary to use the specialised software. Research on the effects of using IWBs (e.g., Higgins, Beauchamp, & Miller, 2007) suggests that many teachers and children are enthusiastic about their use, but empirical evidence of improvements in achievement is less evident to date.

Most virtual manipulatives have been designed for use by individuals or pairs of children, on the assumption that a single computer screen is involved. As interactive whiteboards have become more available in schools, teachers have experimented with software such as virtual manipulatives with the entire class at once. While many IWBs seem to serve a role mostly as a screen (i.e., more 'board' than 'interactive'), some virtual manipulatives lend themselves well to a pedagogy involving children (instead of teachers) using the IWB and class engagement with deciding what manipulations to do and with thinking about the mathematical ideas involved. For that reason, boards are usually mounted low in the classroom, so that young (and short) children can operate them relatively easily, either in a small group or in plenary mode.

An excellent collection of materials of particular value for interactive whiteboard use is provided by Spencer Riley, an innovative UK teacher (Riley, 2012). These contain good instructions for teachers as well as engaging opportunities for classes with an IWB. A good example is shown in Figure 4, designed to support the reading of scales of various kinds, which of course involves a good sense of number in use.

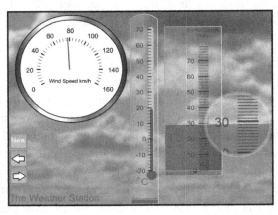

Figure 4. Reading Scales on an interactive whiteboard (Riley, 2012)

Again, it is important to focus students' attention on the key mathematical ideas when using virtual manipulatives of this kind. In this case, the only manipulation involved is to tap the 'new' key to obtain a different set of readings, or the arrow keys to choose a different set of scales. However, a class discussion about how to handle the various markings and the inevitability of approximations being involved whenever measurement is undertaken can be quite productive. Students benefit from the experience of making sure that the markings are well understood on scales and, generally speaking, classes of students will always have some who are misreading the scales, so that fertile discussions will be stimulated by the ICT in this case. The IWB becomes a shared space for whole class discussions.

Rex Boggs, an Australian teacher, maintains a collection of close to 500 digital objects for interactive whiteboards (Boggs, 2012), a very useful resource for teachers. While most of these concern mathematics well beyond the middle years of schooling, many others would be of value to upper primary and middle years teachers, and the classification scheme used on the site makes it relatively easy for teachers to find materials that suit their teaching levels. Similarly, a UK site based at the Keele University (2013) offers many links to resources relevant to interactive whiteboards; some (but not all of these) involve subscriptions.

6 Tablets and Smartphones

The recent popularity of tablets (such as Apple's *iPad* or *iPod Touch*) might seem to give rise to opportunities for students to learn mathematics through the use of ICT. In time, this may be the case, but to date the results are a little disappointing because of limitations of the devices. The most attractive features of virtual manipulatives are the opportunities for students to engage in manipulations of dynamic objects, so the inability of many machines to handle either Java or Flash objects is a severe constraint on the possibilities at present (Kissane, 2011). This might change if more publishers of digital objects agreed to use the new html5 protocols for programming on the web. Already, there are some examples of virtual manipulatives developed for one platform that have been redesigned to work on an *iPad*, but there are only very few of these at present (some are on *Illuminations*, *TeacherLED* and *Interactivate*).

Kissane (2011) noted that a very large proportion of available mathematics applets for the *iPod Touch* used a pedagogy of drill and practice, but there are exceptions exploiting these new technologies for learning purposes. Two examples are shown in Figure 5.

8	6	12	17	7	11	9
	20	1	22	25	10	22
	9	15	25	23	24	19
21	9	19	1	23	25	5	7	9
20	23	13	18	6	14	1	3	16
24	16	22	10	12	3	20	22	24
7	18	16	2	25	19	3	8	14
1	5	22	13	3	4	23	19	21
24	12	14	24	9	14	6	12	10

Figure 5. Kakooma and *Motion Math* on an *iPad*

The game of *Kakooma* involves children touching the (unique) number in each 3 x 3 grid that is the sum of two other numbers in the same grid. In the screen shown, the number in the top grid is 15 (= 6 + 9). Finding this number mentally is often surprisingly difficult, and provides children with a rich opportunity to look for connections between numbers, using both addition and subtraction, and under some time pressure as game times are recorded. The *Motion Math* applet shown in Figure 5 uses the dynamic properties of the device, requiring the user to tilt the tablet so that the numbered ball lands on the correct place on the number line. While each of these mostly provides an opportunity to refine an understanding of numbers and operations, the ICT arguably offers a different learning experience from what is normally available.

7 Representing Numbers on Calculators

Calculators were originally designed to address some of the computational needs of scientists and engineers; hence, early models were referred to as 'scientific' calculators. Recent models have been developed to suit the needs of students learning mathematics, however, although they are still referred to as 'scientific' calculators. Recent changes of importance to the present discussion include functionality to handle various representations of numbers (in particular, fractions and percentages as well as decimals), as well as improved ease of use.

    Despite substantial design changes to support the learning of mathematics, calculators continue to be misunderstood as primarily devices for completing calculations. The decision to include their use in the Singapore Primary Syllabus only for P5 and P6 possibly reflects such a view, as several entries clearly refer to calculations "without using calculators". Research and developmental work many years ago by Shuard et al., (1991) in the UK and by Groves and Stacey (1998) in Australia suggested that calculators can be helpfully used for learning by children even in the early years of primary school, however. Teachers of younger children will find this work of particular interest, with ample evidence available that allowing children to use calculators will not be harmful to their learning of mathematical ideas and indeed can enhance the learning considerably, especially when care is taken to encourage productive use.

    A modern calculator, such as the CASIO fx-95 SG PLUS, offers older children the opportunity to explore relationships among numbers in a powerful way. The screens in Figure 6 show that pupils can see for themselves through their own activity that fractions and decimals are merely different representations of the same number, through using the calculator key designed to switch between fractions and decimals. (In these, and other screens, the first line shows what is entered into the calculator, while the second line shows the result.) The calculator provides immediate, powerful and frequently interesting feedback.

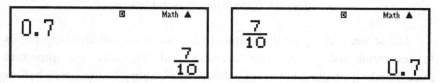

Figure 6. Representing a number as a fraction and as a decimal

    While it is fundamental that seven tenths is represented by 0.7, as this is the key idea of place value, children using a calculator can be surprised that some fractions are not represented by the calculator in a way that they might at first expect, as shown in Figure 7.

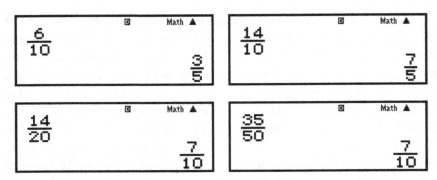

Figure 7. Equivalent fractions on a calculator

Equipped with such a device, students can see for themselves that many different fractions are represented by the same decimal, which is the critical understanding of the important concept of equivalent fractions.

As well as representing fractions as decimals, the calculator has a key for changing representations from improper to proper fractions, as illustrated in Figure 8.

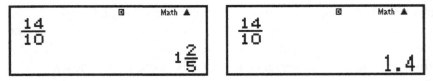

Figure 8. Representing improper fractions

The screens in Figure 9 illustrate that calculators can also help pupils to see relationships between division and fractions, an important understanding of the concept of a fraction. Contrary to the naive beliefs of some children, a calculator makes it quite clear that it is perfectly acceptable to divide a smaller number by a larger number, and helps a user to see that the result is a fraction. In a similar way, a calculator can help pupils understand that it is also acceptable to subtract a larger number from a smaller number leading to a need to deal with negative numbers: The same kind of calculator can be used by pupils to understand relationships among percentages, fractions and decimals.

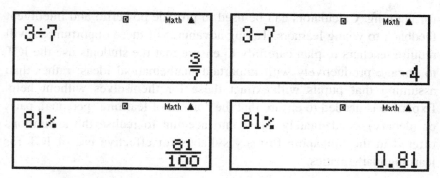

Figure 9. Using a calculator to explore representations of numbers

These various examples illustrate a general property of a modern calculator: the many different representations of numbers are faithfully represented by the calculator and can be readily manipulated by pupils to help them to see the connections between them. These kinds of capabilities were not available on early scientific calculators (which were not intended primarily to support the learning of mathematics, but the computational needs of scientists and engineers). The calculator is best regarded as a responsive device, incorporating the real number system with it, from which pupils can learn about the number line from many different perspectives through their own explorations. Its significance for education comes from these kinds of learning opportunities, and not from its computational capabilities. Ideas of these kinds are explored in some detail in Kissane & Kemp (2013), especially (but not only) for older students.

8 Conclusion

Much is to be gained by both teachers and pupils in exploring the potential of ICT to provide different learning opportunities in mathematics. Virtual manipulatives especially offer fresh ways to engage pupils in thinking about mathematics to support the development of their understanding of key concepts and relationships. There are very many excellent sources of these on the Internet at minimal cost. Newer technologies of tablets and interactive whiteboards offer new possibilities for learning. Calculators can be used to provide powerful and intelligent feedback to young learners. Taking advantage of these opportunities will require teachers to plan carefully to ensure that the students use the ICT to engage productively with important mathematical ideas, rather than assuming that pupils will extract these for themselves without help. Expertise is needed to ensure that the context of learning (personal, pairs or plenary) is adequately taken into account to realise the aspirations offered in the Singapore Primary syllabus for effective use of ICT for learning mathematics.

References

Becta (2009). *Primary mathematics with ICT: A pupil's entitlement to ICT in primary mathematics*. Retrieved 6 March 2014 from
http://www.nationalstemcentre.org.uk/elibrary/resource/4537/primary-mathematics-with-ict-a-pupil-s-entitlement-to-ict-in-primary-mathematics

Boggs, R. (2012) *Interactive mathematics classroom*. Retrieved 6 March 2014 from
http://interactivemaths.net/

Cambridge University Press (2012). *HOTmaths*. Retrieved 6 March 2014 from
http://www.HOTmaths.com.au/

Center for Implementing Technology in Education (2012). *Learning mathematics with virtual manipulatives*. Retrieved 6 March 2014 from
http://www.cited.org/index.aspx?

Education Services Australia (2010). *Maths300*. Retrieved 6 March 2014 from
http://www.maths300.esa.edu.au/

Groves, S., & Stacey, K. (1998). Calculators in primary mathematics: Exploring number before teaching algorithms. *The teaching and learning of algorithms in school mathematics: 1998 Yearbook of the National Council of Teachers of Mathematics* (pp. 120-129). Reston, VA: NCTM.

Higgins, S., Beauchamp, G., & Miller, D. (2007) Reviewing the literature on interactive whiteboards. *Learning, Media and Technology, 32(3)*, 213-225.

Keele University (2013). *Interactive whiteboard mathematics*. Retrieved 5 March 2014 from http://iwbmaths.co.uk

Kissane, B. (2009). What does the internet offer for students? In C. Hurst, M. Kemp, B. Kissane, L. Sparrow & T. Spencer (Eds.) *Mathematics: It's Mine*: Proceedings of the 22nd Biennial Conference of the Australian Association of Mathematics Teachers. (pp. 135–144) Adelaide: Australian Association of Mathematics Teachers. Retrieved 6 December 2012 from http://researchrepository.murdoch.edu.au/6243/

Kissane, B. (2011). The iPod touch and mathematics education. In J. Clark, B. Kissane, J. Mousley, T. Spencer & S. Thornton (Eds.) *Mathematics: Traditions and [New] practices:* (Proceedings of the joint 32nd annual conference of the Mathematics Education Research Group of Australasia and 23rd biennial conference of the Australian Association of Mathematics Teachers, pp. 932-940). Adelaide: AAMT/MERGA. Retrieved 6 December 2012 from http://researchrepository.murdoch.edu.au/6202/

Kissane, B. (2013a). *Learning mathematics and the internet*. Retrieved 6 March 2013 from http://wwwstaff.murdoch.edu.au/~kissane

Kissane, B. (2013b). *Interactive mathematics on the internet*. Retrieved 6 January 2014 from http://wwwstaff.murdoch.edu.au/~kissane/pd/javamaths.htm

Kissane, B., & Kemp, M. (2013) *Learning mathematics with an advanced scientific calculator*. Tokyo: CASIO.

Ministry of Education (2012). *Primary mathematics teaching and learning syllabus*. Singapore: Author.

Moyer, P. S., Bolyard, J. J., & Spikell, M. A. (2002). What are virtual manipulatives? *Teaching Children Mathematics*, 8, 372-377. Retrieved 6 December 2012 from http://www.grsc.k12.ar.us/mathresources/Instruction/Manipulatives/Virtual Manipulatives.pdf

National Council of Teachers of Mathematics (2012). *Illuminations*. Retrieved 6 December 2013 from http://illuminations.nctm.org/

Pierce, R. (2013). *Maths is fun - Maths resources*. Retrieved 5 November 2013 from http://www.mathsisfun.com/

Riley, S. (2012). *TeacherLED: Interactive whiteboard resources for teachers*. Retrieved 6 March 2014 from http://www.teacherled.com/

Shuard, H., Walsh, A., Goodwin, J., & Worcester, V. (1991) *Calculators, children and mathematics*. London: Simon & Schuster.

Shodor Educational Foundation (2012). *Project interactivate*. Retrieved 6 march 2013 from http://www.shodor.org/interactivate/

University of Cambridge (2012). *Nrich*. Retrieved 6 March 2014 from http://nrich.maths.org/frontpage

University of Cambridge (2013). *Introduction to fractions*. Retrieved 4 June 2013 from http://nrich.maths.org/5540

Utah State University (2010). *National library of virtual manipulatives*. Retrieved 6 March 2014 from http://nlvm.usu.edu/en/nav/vlibrary.html

Wheatley, G. R. (1992). The role of reflection in mathematics learning. *Educational Studies in Mathematics, 23*(5), 529-541.

Chapter 14

Learning Algebra and Geometry through ICT

Marian KEMP

Information and communications technology (ICT) offers different kinds of experiences for learning mathematics from those found in textbooks and traditional teaching in the classroom, for primary and secondary pupils. This chapter examines how features of ICT can be used to help create mathematical meaning for pupils, inside and outside the classroom in the areas of algebra and geometry for secondary pupils. The initial incorporation of the use of ICT may bring practical challenges and technical problems for teachers. In addition, the use of ICT requires some change in the roles by both pupils and teachers. Teachers will need to be mindful of potential pitfalls and be ready to think about when the use of ICT is sensible. This chapter will outline some advice about how teachers can use ICT to their advantage and for the pupils' enjoyment and learning.

1 Introduction

This chapter focuses on how teachers and pupils can use ICT in the classroom to help pupils develop an understanding of mathematical concepts and ideas, especially those in algebra and geometry. The work of Drent and Meelissen (2008) in Holland with teacher educators identified three objectives for the study of the use of ICT in education: "The use of ICT as object of study; the use of ICT as aspect of discipline or profession; and the use of ICT as medium for teaching and learning" (p. 187). All three objectives are relevant in secondary schools but the focus here is on the use of ICT as a medium for teaching and learning.

The British Educational Communications and Technology Agency (Becta, 2003) produced a report about ICT and motivation. This report was one of a series that analyses the available evidence on the use of ICT in education. The report concluded that ICT can stimulate, motivate and spark students' appetites for learning and helps to create a culture of success.

In the Singapore secondary mathematics syllabus (Ministry of Education, 2012) it is explicitly stated that:

Teaching (*of mathematics*) should connect learning to the real world, harness ICT tools and emphasise 21st century competencies. (p. 21)

The syllabus elaborates further the harnessing of ICT tools as follows:

Teachers should consider the affordances of ICT to help students learn. ICT tools can help students understand mathematical concepts through visualisations, simulations and representations. They can also support exploration and experimentation and extend the range of problems accessible to students. The ability to use ICT tools is part of the 21st century competencies. (p. 22)

In this chapter reference to ICT will include the use of computers while connected to the Internet, perhaps using an interactive whiteboard for groups or whole class interaction; previously downloaded materials from the Internet; appropriate software provided on pupil's computers and a range of capabilities afforded on modern day calculators. All of these provide opportunities for pupils to become engaged with the mathematics in different ways. In this chapter specific suggestions will be restricted to the areas of algebra and geometry for secondary school pupils.

Teachers are busy, professional people with an extensive syllabus to follow, set to achieve the standards of mathematical knowledge appropriate for pupils in the various grades. Unless the use of ICT is of value to the teachers and their pupils, and is integrated into the curriculum then its use is questionable.

The Singapore Syllabus does not just concentrate on content knowledge. In addition to the describing the development of concepts skills and processes it is serious about developing pupils' ability to (i) undertake mathematical reasoning, communication and connections (ii) use thinking skills and heuristics to solve problems and (iii) develop problem solving skills to tackle problems including real-world problems (MOE, p. 14). Therefore in considering the use of ICT in the classroom it is important to bear this in mind in the selection of lessons and activities.

All of the websites cited in this chapter have a range of offerings which include some or all of: interactive activities for pupils - perhaps graphing or space orientated; lesson plans for teachers to carry out with or without ICT; worked examples with 'voice overs'; exercises based on the examples; mental computations; games; articles about real life mathematics; 'real world' problems and problem solving skills as well as activities to develop mathematical thinking, reasoning and proof.

2 Hand Held Technology

Over the last few decades there have been huge changes in the access to technology in schools and at home. In the early 1970s, four function and scientific calculators became very popular with mathematicians and engineers and reached some schools but they were very expensive. Throughout the 1970s the only access many pupils had to computers was through the computer cards used to write programs for computers that may have been on or off site. Then there was the possibility of a single computer for demonstration purposes in the classroom and a few computer laboratories that had to be booked for classes in advance. In some underdeveloped countries this is still the case.

Following on from this situation, in the mid-1990s, handheld technology HHT in the form of graphics and symbolic calculators became widespread in schools in developed countries. Trouche and Drijvers (2010) have suggested that this technology became so widespread in part because implementation of technology in the classroom was so difficult. The HHT is personal which allows pupils to try out different things and make errors (Ruthven, 1990), and it is always available (Bradley, Kemp & Kissane, 1994; Lagrange, 1999).

Graphics calculators come equipped with the software that enables pupils to see the connections between functions and their graphs. In addition, pupils can learn about and interact with other algebraic concepts including recursion, series and sequences. Graphics calculators can be used to solve equations and simultaneous equations in a variety of ways; equation solvers, using matrices and graphical methods.

Most modern graphics calculators have integrated dynamic geometry software systems. Pupils can learn about the geometric properties of shapes and develop understandings of geometric relationships such as angle properties of a circle and transformations including reflection, rotation and translation. Indeed, Drijvers and Weigland (2010) noted that the symbolic calculators with their inbuilt dynamic geometry systems and algorithmic operational systems would enable pupils to concentrate more on concept development and problem solving and modelling competencies.

Current scientific calculators have a range of capabilities. These have changed significantly over the last twenty years. Instead of being regarded as merely computational tools, they can be used to help pupils understand mathematical concepts. For example, one advanced scientific calculator allows letters to be used to represent equations, which can then be solved numerically. The advanced scientific calculators include capabilities that are just short of the graphing features of graphics calculators.

3 Laptops and Tablets

It is only recently that access to ICT has changed with pupils having laptops and other electronic devices for use at any time. Unfortunately, at the moment, many tablets cannot handle the use of *Java* or *Flash* objects that are often used in the interactive activities offered on the Internet (Kissane, 2011). This means that pupils need to have Java and Flash installed and activated on their laptops to access these activities in and outside of the classroom.

When pupils were in laboratories it was very much an individualised culture with each pupil at their own station carrying out (or not!) the instructions from the teacher. In describing the use of ICT in current classrooms, Sutherland et al (2004, p. 418) emphasise that we should move towards a balance of individual, group work and whole class work and that we should develop classrooms where students are confident to discuss their work.

4 What Do Pupils Get From ICT?

A very useful summary of the opportunities related to ICT appears in a British document produced by Becta (2009) (a government agency that has since been disbanded as a result of structural governmental changes). The material has been reworked by the National Strategies and is now available in the STEM e-library. This document identified and exemplified six major opportunities for pupils to benefit from the use of ICT in mathematics, describing them as 'entitlements', each of which is elaborated a little below:

Learning from feedback
ICT provides fast reliable feedback that is non-judgemental. Pupils can make conjectures and test their ideas.

Observing patterns
The computers can produce many patterns for learners in a short time. This is very useful in algebra to enable pupils to test and generalise algebraic relationships. In addition, in geometry pupils need to test and develop proofs of geometric relationships.

Seeing connections
ICT tools provide ways for pupils to see connections between mathematical ideas, especially the so-called 'rule of three' concerned with representing functions with symbols, graphs and tables of values.

Computer software, spread sheets and graphics and symbolic calculators provide quick and dynamic ways for these kinds of connections to be seen and used to understand relationships, through switching from one representation to another.

Developing visual imagery
Especially in recent years, ICT devices such as computers and tablets have provided new visual opportunities for learners, so that learning mathematics is no longer restricted as much to symbolic and numerical representations. The fresh opportunities created by ICT go beyond providing visual images, however, as the images can often now be generated and manipulated by pupils.

Exploring data
In this information age, most data are stored in computers, so that they can be analysed, represented and used efficiently. Similarly for pupils, when data are involved, the use of computers allows for real data to be used rather than relying on data cleansed for classroom use, and for a variety of representations and analyses to be efficiently undertaken.

'Teaching' the computer
Modern user-friendly ICT devices have been designed to allow us to interact easily with devices (e.g., using a mouse, a stylus or a touch screen in a software environment). To an extent, this has hidden the detail with which such interactions must be designed and described. Engaging pupils in 'teaching' a computer through writing instructions and designing algorithms creates an opportunity for them to understand the precision with which commands need to be communicated.

The Becta document provides opportunities and gives many examples that Singaporean teachers will find informative and helpful.

5 The Roles and Responsibilities of the Teacher

There is inevitably some shift in pedagogy and associated teaching practices when ICT is introduced into the classroom. Hayes' (2007) report on ICT in Australian classrooms noted that: "successful integration of ICT requires fundamental shifts in the core activities of the schools. These shifts include new teaching" (p. 385). Hayes observed that teachers' success in integrating ICT was dependent on the availability of time to think about pedagogical practices and to work with colleagues in a community of practice to develop programs. With the introduction of ICT to the classroom the teacher remains critical to the learning of the pupils – the technology does not do the teaching (Drijvers and Weigland, 2010). It is the teacher who needs to provide a learning situation to motivate and engage the pupils (Sutherland et al., 2004). Not all lessons will use ICT; it is important that both pupils and teachers can decide when it is suitable, and when it is not.

In order to provide a suitable learning environment for pupils it is expected that the teacher will need some professional development. This may include professional development in using graphics calculators, if these are allowed in the schools, or intensive training in Internet use. Unfortunately the Australian study (Hayes, 2007) found that teachers tend to continue to teach in the way that they had always taught despite access to ICT. The extent of access to good technical support makes a big difference. For example one teacher in Hayes' study reported that:

> There's a constant maintenance problem because things go wrong-just bits and pieces. But you have to be able to resolve it yourself. You've really got to become adept at sort of understanding your own computers. And you tend to if you have them long enough (Hayes, 2007, p. 394).

It is easy to give up on ICT if the technology does not work; very reasonably teachers do not want to look ineffectual in their own classrooms.

Sutherland et al. (2004) found that teachers new to using ICT tend to focus on the ICT discourse rather than on the subject focussed language.

This is when teachers are uncertain and lack confidence about the ICT they are using. Therefore, professional development is very important whether it is in the use of calculators, computers, tablets, iPads, laptops or interactive whiteboards. Collaboration between teachers and attendance at conferences may help.

6 Interactive Whiteboards

Interactive whiteboards (IWB) can be used in a variety of ways to suit the classroom activities and pupils. When the boards are connected to the Internet they can be used to access software programs through the computer with which they are connected. Teachers can use their chosen software for demonstration purposes, to motivate pupils and to engage them in the learning process. Pupils may access appropriate software themselves, in pairs or small groups. In particular they can be used to encourage discussion using mathematical language. The feedback from the software is non-judgemental which encourages pupils to 'try again'. For example, using *GeoGebra* on an interactive whiteboard provides a medium for students to estimate and draw angles of a particular size and then measure them.

British Educational Communications and Technology Agency (Becta, 2009) analysed the literature on interactive whiteboards. Their analysis indicated that interactive whiteboards can have positive effects on teaching and learning in general, and provide benefits for teachers and for students.

One of the studies included in the Becta analysis identified three levels of whiteboard use:

- to increase efficiency, enabling teachers to draw upon a variety of ICT-based resources without disruption or loss of pace
- to extend learning, using more engaging materials to explain concepts
- to transform learning, creating new learning styles stimulated by interaction with the whiteboard (Glover & Miller, 2001).

Glover et al (2007) describe the latter learning styles as *Supported didactic, Interactive* and *Enhanced interactive*. There is a tendency for the focus in the first two to revert to the teacher but on the third one the teacher "orchestrates full dialogue with and between pupils and cooperates with other teachers. Pupils are seen as 'equals learning together'" (p. 12). The teachers combine different modes of learning using verbal, visual and kinaesthetic approaches.

There are many resources on the Internet that teachers can access without cost, some of which are listed in the Appendix. There are other good ones that require a cost but they are not included.

7 Resources for Algebra

In the Singapore Curriculum it is expected that all pupils will become very familiar with symbolic representation and formulae. Pupils need to be able to understand what it means to represent a relationship in an equation and to solve linear, quadratic and cubic equations with integer and non-integer coefficients.

7.1 *Linear equations*

In grades 6 to 8 to help pupils understand the concepts of formulae they might use the National Library of Virtual Manipulatives *(NLVM): Algebra Beam Balance* and *Algebra Negative Beam Balance*. Figure 1 shows the beginning of a pupil solving an equation, balancing the beam as they go.

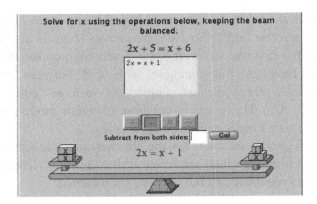

Figure 1. From *NLVM* solving $2x+5 = x + 6$

7.2 *Quadratic equations*

Teachers can help pupils see different representations: symbolic solutions using factorisation, the quadratic formula or completing the square; as well as tables and graphical representations.

A scientific calculator can be used in several ways to solve the equation $x^2 - 3x - 2 = 0$. Firstly, pupils would normally attempt to factorise the expression. Since the expression cannot be factorised the pupils would then use the quadratic formula or completing the square. However, now many advanced scientific calculators include a table function that can be used to solve a range of equations through iteration. This is particularly useful for solving quadratic or cubic equations that do not have integer solutions and so will not factorise.

In Figure 2 the *NLVM* provides pupils with the opportunity to see graphically the approximate solutions to a quadratic equation, by examining the roots of the corresponding function. The trace button can be used to find the solutions to $x^2 - 3x - 2 = 0$ are $x = -0.55$ and $x = 3.55$ with $f(x) = -0.05$. This does not give an exact answer but is quite close. The window can be changed to get further accuracy when zoomed in.

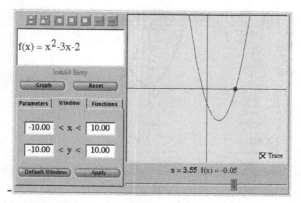

Figure 2. Using the function grapher in the NLVM

7.3 *Investigating functions*

When teachers and pupils use graphing software or a graphics calculator to see families of functions and transformation of functions they are able to see them visually. Figure 3 shows the vertical transformation of $y = x$, to the functions $y = x + 2$ and $y = x - 3$

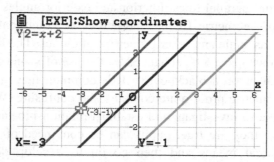

Figure 3. Transforming a linear function using a graphics calculator CASIO *fx*-CG 20

As pupils begin to learn about other graphs including quadratics, cubics, exponentials, trigonometry and conics they can become efficient in understanding the changes that different parameters cause. Figure 4 shows the horizontal translation of $y = x^2$ two places to the right by graphing $y = (x - 2)^2$.

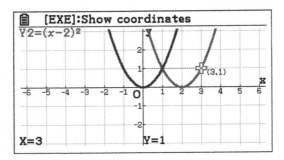

Figure 4. Transforming a quadratic function using a graphics calculator CASIO *fx*-CG 20

Pupils can experiment in a range of graphing software programs to see the change in effect of modifying the parameters on the shape of the graph.

8 Resources for Geometry

Geometric properties and relationships are an important part of the Singapore mathematics curriculum. Pupils learn about angle relationships and parallel lines; interior and exterior angles of polygons; construction of geometric figures; properties of congruent and similar shapes as well as angle and other properties of circles. The learning of all of these relationships can be enhanced by the use of visual aids; whether they are digital or on a blackboard. It is common that pupils find the concept of geometric proof quite complex and confusing. Whilst the use of digital media to investigate and establish relationships and properties does not prove they are true, pupils can be convinced of their validity through empirical means and then go on to develop deductive and inductive proofs.

Dynamic geometry systems (DGS) are tools that have been available since about 1990, offering experiences to students and teachers that are not readily available in other ways. Internationally, the best commercial examples of software of this kind have been *Cabri Geometry* (http://www.cabri.com) and Key Curriculum Press's *The Geometer's Sketchpad* (http://www.dynamicgeometry.com), each of which requires the purchase of licenses to use the software. In recent years, *GeoGebra* (http://www.geogebra.org) has been developed as a freely available alternative to commercial software. These systems are best regarded as 'tool' software, in the sense that they allow the user (both pupil and teacher) to use tools to *do* mathematical things, rather than 'tutor' software that is directed at 'teaching' students directly, 'demonstrating' or 'explaining' things to students, 'testing' students or getting them to 'practise' some aspects of mathematics already learned.

Pupils can learn individually, in groups or whole class situations by interacting with a DGS. In addition, some of the Internet sites such as *WaldoMaths*, the *National Library of Virtual Manipulatives*, *Geometry Open Book* and *Inter*activate provide some facilities to investigate geometric properties. The examples in Figure 5 and Figure 6 have used the *GeoGebra* software. The tools shown as buttons on the toolbar allow pupils to draw objects of various kinds and to manipulate and measure them in various ways in order to explore their properties.

In the example shown in Figure 5, the *Polygon* tool has been used to draw a rectangle. Lengths, angles and area can be found using the tools in the toolbar. In Figure 6, the vertices A and D have been moved to form a parallelogram on an equal base to the rectangle and between the same parallels, the area is unchanged and the sum of the angles is still 360° but the slanting sides are longer. Pupils can investigate properties and make conjectures and try to prove them for themselves. The software helps students to develop ideas but it does not provide formal proofs. Instead, pupils need to develop their own arguments to explain and to justify their conclusions.

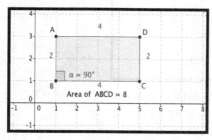

Figure 5. Rectangle ABCD

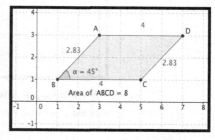

Figure 6. Parallelogram ABCD

Pupils can investigate the effects of transformations such as translation, reflection, rotation and shear on different shapes. Figure 7 shows the reflection of triangle ABC in the given line where pupils can see that each point on the new triangle A'B'C' is equidistant from the line of reflection as ABC and that the shape is invariant. Pupils can try other shapes and other axes of reflection.

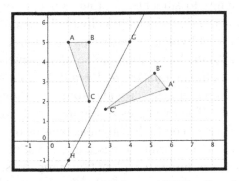

Figure 7. Triangle ABC and its image A'B'C' reflected in HG

When pupils investigate rotation such as in Figure 8 they can vary the shape and size of the object and the centre and angle of rotation. A slider can be inserted to change the angle if desired.

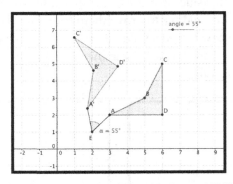

Figure 8. Quadrilateral ABCD and its image A'B'C'D'
after anticlockwise rotation of 55° about (2, 1)

The website *WaldoMaths* has an extensive area devoted to geometry. There are many problems for pupils to attempt. In particular pupils are able to explore the circle properties and come to the circle theorems. The applets use JAVA to make them interactive. In Figure 9, as the vertices of the quadrilateral are rotated around the circle the angles change. Other applets engage pupils interest in the other circle properties.

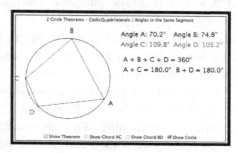

2 Circle Theorems - CyclicQuadrilaterals / Angles in the Same Segment

Angle A: 70.2°   Angle B: 74.8°
Angle C: 109.8° Angle D: 105.2°

A + B + C + D = 360°
A + C = 180.0°   B + D = 180.0°

☐ Show Theorem   ☐ Show Chord AC   ☐ Show Chord BD   ☑ Show Circle

Figure 9. Circle theorems applet from WaldoMath

In addition to the computer software, which can be used on laptops or personal computers, pupils may have access to a graphics calculator. These have the software for learning mathematics already installed on them. Many now have a dynamic geometry package that can be used in similar ways to the software on computers.

9 ICT Resources and the Curriculum

Table 1 to Table 6 outline the Singapore Syllabus (MOE, 2012) for secondary pupils with some suggestions as to the websites and appropriate materials that teachers might use for algebra and geometry. Some of the learning opportunities target pupils and others provide advice for teachers.

These resources are by no means exhaustive but are intended as a guide to provide some information for teachers. The suggestions are put alongside the main topics; the details are in the syllabus. The websites contain material teachers can adapt to use with pupils in the different levels as the resources span the levels. Some specific resources are given in the tables; teachers will find many more in the list of websites that are accessible without payment at the end of the chapter. It is important to note that there may be changes to these resources as the authors make adjustments. On the websites there is some guidance about the grades, namely 6-8 and 9-12, but there is considerable overlap around the middle. The websites named in the tables give interactive resources for pupils to use, some lesson plans or guidance for teachers and some worksheets.

As the students mature and are able to engage in abstract thought they will be able to tackle different kinds of tasks and move more to understanding of proof in both algebra and geometry. In Secondary One and Two it might be expected that pupils would need more guidance by their teacher; there might be more "whole class activity" with the teacher leading the discussion and questioning. This approach could be easily facilitated by using an interactive whiteboard. As the pupils mature and become more independent they could work in pairs and small groups around the IWB or a laptop, developing their own discussions and questioning.

It would also be expected that pupils could approach more abstract concepts through using ICT where appropriate. Even for older pupils the visualisation helps them to develop those concepts.

Table 1

Examples of ICT for Algebra Secondary One

Topic	Resource	Appropriate ICT tools
Algebraic representation and formulae	***NLVM*** ***NRICH***	*Function machine:* Looking for patterns to identify relationships *Interactive number patterns 1 and 2:* Pupils could work in pairs to identify the patterns and find the formulae *Reasonable sums:* This builds on consecutive sums to develop generalisations *Sequences and series & More sequences and series:* Pupils build on with arrays of triangle

		and odd numbers, developing expressions for general terms *Harmonic triangles:* pupils explore unit fractions in the harmonic triangle
Algebraic manipulation	**Illuminations** **NRICH**	*Algebra tiles:* Visual development of concepts of factorisation and expansion of expressions through virtual manipulation of "tiles" *Algebra:* Manipulating algebraic expressions/formulae
Functions and graphs	**NLVM** **Interactivate** **NRICH**	*Function machine:* Given a set of numbers *(x)* and gradually numbers for (*y)* pupil*s* look for patterns to find the relationship between the variables *Line plotter:* Pupils construct a line with a given point and gradient and get immediate feedback about correctness of their plots *Grapher:* Pupils plot linear, quadratic, trig and exponential graphs and explore the relationships using different parameters *Slope slider:* An activity where pupils explore the relationship between the y-intercept, the gradient and the equation of the linear function using a slider to change the parameters *Parallel & perpendicular lines:* Pupils explore the characteristics of the equations that define parallel and perpendicular lines
Solutions of equations and inequalities	**NLVM** **Interactivate**	*Algebra balance scales:* Pupils solve equations of the form $Ax + B = Cx + D$ (with integer solutions), using a balance and virtual (drag and drop) movement of the expressions on both sides *Connect 4:* A game for 2 players who have to solve linear equations to progress and win

	Illuminations	*Algebra quiz:* Timed solution of linear and quadratic equations with 4 levels of difficulty *Equation solver:* Uses identity properties in solving linear equation with 3 levels of difficulty, identifying Identity Properties
	Maths is Fun	*Pan balance-expressions:* Algebra topic

Suggestions made in the websites show some overlap for the secondary years.

Table 2

Examples of ICT for Algebra Secondary Two

Topic	Resource	Appropriate ICT tools
Algebraic manipulation	*NLVM* *Maths Online* *NRICH*	*Algebra tiles:* Factoring and expanding polynomials *Variables, terms, formulae, identities* *Generalisation package*
Functions and graphs	*NLVM*	*Grapher:* Pupils can plot up to three functions on the axes. Linear, quadratic, trig and exponential graphs can be plotted.
	Interactivate *NRICH*	*Function transformations:* Pupils can explore how simple transformations can affect the graph of a function *Slope Slider* *Quadratic functions and More quadratic transformations:* Pupils may graph and compare their chosen functions or guess someone else's *Parabolic patterns:* Pupils compare graphs in the form $a(x-b)^2$ *Up and across:* This involves guessing the paths (loci) taken by points. . *Walk and ride:* In this one pupils investigate simple related rates

Solutions of equations	**Maths Online** **Skoool**	Coordinates and graphs Equations

Table 3

Examples of ICT for Algebra Secondary Three/Four

Topic	Resource	Appropriate ICT tools
Functions and graphs	**GeoGebra** **Graphics** **calculators** **NRICH**	*Generalising package has a range of activities*
Solutions of equations and inequalities	**GeoGebra** **Graphics** **calculators** **NRICH**	*Generalising package has a range of activities*

Table 4

Examples of ICT for Geometry for Secondary One

Topic	Resource	Appropriate ICT tools
Angles, triangles and polygons	**WaldoMaths** **Interactive** **GeoGebra**	*Angles* *Shape, space and measure:* vertically opposite angles, alternate, corresponding and adjacent angles angle sum, straight lines and exterior angles Shapes, and angles can be constructed and measured

One of the advantages of using dynamic geometry systems is that pupils can move the shapes or angles and place them on others. Although the systems do not prove theorems pupils can see the relationships.

Table 5

Examples of ICT for Geometry for Secondary Two

Topic	Resource	Appropriate ICT tools
Congruence and similarity		Sites below include rotation, translation, reflection, enlargement and symmetry
	NLVM	Congruent triangles, Transformations
	Illuminations	Congruence explorer
	Interactivate	Transmographer, transmographer 2, 3D transmographer
	WaldoMaths	Shape, space and measure 11-16
Pythagoras Theorem	*NLVM*	Pythagoras
	Interactivate	Pythagoras, four levels
	WaldoMaths	Shape, space and measure: The Pythagoras rule

These websites provide plenty of experience with transformations where visual experience is so important.

Table 6

Examples of ICT for Geometry for Secondary Three/Four

Topic	Resource	Appropriate ICT tools
Congruence and similarity	**NLVM**	Platonic solids, congruent triangles,
	Illuminations	transformations
	Interactions	Congruence explorer
	WaldoMaths	Transmographer, Transmographer 2, 3D Transmographer
		Shape, space and measure 11-16
Properties of circles	*WaldoMaths*	Shape, space and measure: cyclic quadrilaterals, angles in the same segment,

	angle in a semicircle, the alternate segment theorem, tangent and radius, tangents from a point, arc lengths, areas of sectors and segments

These sites help pupils to develop an understanding of the relationships; pupils need to develop their ability to explain and justify their findings leading into formal proofs.

References

Becta. (2003). *What the research says about ICT and motivation.* Retrieved March 4, 2014, from http://partners.becta.org.uk/index.php?section=rh&rid=13660

Becta. (2009). *Secondary Mathematics with ICT: A pupil's entitlement to ICT in secondary mathematics.* Retrieved 30 May 2013, from http://www.nationalstemcentre.org.uk/elibrary/resource/4538/secondary-mathematics-with-ict-a-pupil-s-entitlement-to-ict-in-secondary-mathematics

Bradley, J., Kemp, M., & Kissane, B. (1994) Graphics calculators: A (brief) case of technology. *Australian Senior Mathematics Journal,* 8 (2), 23-30.

Drent, M., & Meelissen, M. (2008). Which factors obstruct or simulate teacher educators to use ICT Innovatively? *Computers & Education, 51,* 187-199.

Drijvers, P., & Weigland, H-G. (2010). The role of handheld technology in the mathematics classroom. *ZDM Mathematics Education, 42,* 665-666.

Glover, D., Miller, D., Averis, d., & Door, V. (2007). The evolution of an effective pedagogy for teachers using the interactive whiteboard in mathematics and modern languages: an empirical analysis from the secondary sector. *Learning, Media and Technology, 30*(1) 5-20, DOI: 10.1080/17439880601141146

Glover, D., & Miller, D. (2001). Running with technology: the pedagogic impact of the large scale introduction of interactive whiteboards in one secondary school. *Journal of Information Technology for Teacher Education, 10*(30), 257-275

Hayes, D. (2007). ICT and learning: Lessons from Australian classrooms. *Computers and Education, 49*(2), 385-395.

Kissane, B. (2011). The iPod touch and mathematics education. In J. Clark, B. Kissane, J. Mousley, T. Spencer & S. Thornton (Eds.) *Mathematics: Traditions and [New] Practices* (Proceedings of the joint 32nd annual conference of the Mathematics Education Research Group of Australasia and 23rd biennial conference of the Australian Association of Mathematics Teachers, pp. 932-940). Adelaide: AAMT/MERGA. [Accessible via http://researchrepository.murdoch.edu.au/6202/]

Lagrange, J-B. (1999). Complex calculators in the classroom: Theoretical and practical reflections on teaching pre-calculus. *International Journal of Computers for Mathematical Learning, 4*, 51-81.

Ministry of Education (2012). *O-level mathematics teaching and learning syllabus.* Singapore: Author.

Ruthven, K. (1990). The influence of graphic calculator use on translation from graphic to symbolic forms. *Educational Studies in Mathematics*, 21(5), 431-450

Sutherland, R., Armstrong, V., Barnes, S., Brawn, R., Breeze, N., Gall, M., Mathewman, S., Olivero, F., Taylor, A., Triggs, P., Wishart, J., & John, P. (2004) Transforming teaching and learning: embedding ICT into everyday classroom practices. *Journal of Computer Assisted Learning, 20*, 413-425.

Trouche, L., & Drijvers, P. (2010). Handheld technology for future mathematics education: flashback to the future. *ZDM Mathematics Education, 42*, 667-681.

Appendix

ICT Resources for Algebra and Geometry

Some free sources available for Algebra and Geometry include:

Core Math Tools (http://www.nctm.org/resources/content.aspx?id=32702)
This suite of downloadable *Core Math Tools* from the NCTM in USA includes a *Computer Algebra System (CAS)* tool that allows for graphs and tables to be generated and algebraic expressions used to explore algebra. It also includes an interactive *Geometry* tool that allows for a wide range of geometric operations to be undertaken.

GeoGebra (http://www.geogebra.org)
GeoGebra is available for the study of algebra and geometry. A collection of web-based applets for New South Wales teachers made with *GeoGebra* is available at *The Geometer's Warehouse*.
(http://lrrpublic.cli.det.nsw.edu.au/lrrSecure/Sites/Web/geometer/index.html)

Illuminations (http://illuminations.nctm.org)
This large site was constructed by the National Council of Teachers of Mathematics in the USA, and provides a wide variety of lesson materials and activities. For example: Algebra: *Algebra Tiles, Function Matching, Pan Balance – Expressions*. Geometry: *Angle Sums, Congruence Theorems, Diagonals to Quadrilaterals, Triangle Classification, Parallelogram Exploration Tool*.

Inter*activate* (http://www.shodor.org/interactivate/)
Shodor's collection contains detailed information for teachers as well as interactive Java applets for individual or classroom use. For example: Algebra: *Graph sketcher, Equation solver, Graphit, Slope slider, Linear inequalities, Function flyer*. Geometry: *Squaring the triangle, Transmographers, Angles*.

National Library of Virtual Manipulatives (http://nlvm.usu.edu)
This large popular US collection offers Java applets across several content areas and ages. For example: Algebra: *Algebra Tiles, Grapher, Algebra Balance Scales – negatives,*

Block Patterns, Function Machine, Line Plotter, Function Transformations. Geometry: *Transformations (Reflection, rotation, translation, composition, dilation), Congruent Triangles.*

NRICH (http://nrich.maths.org)
The NRICH site from the UK provides a variety of materials for learners across a wide range. For example: Algebra: The downloadable *Generalising* package at http://nrich.maths.org/5853 is useful where the emphasis is on developing pupil's mathematical thinking through a range of activities. Live problems provide pupils the opportunity to send in their solutions. Geometry: Try the downloadable *Visualising* package at http://nrich.maths.org/5678

iPads/iPods
These very popular tablets suffer a severe limitation that they cannot use Java and Flash applets. For example: Algebra: *Algebra Touch, MyScript Calculator, Slope Slider, Math Flyer, GraphCalc.* Geometry: *Sketchpad Explorer.*

IWBmaths (http://iwbmaths.co.uk)
A UK site based at the University of Keele with many links to resources relevant to interactive whiteboards. Some (but not all) involve subscriptions.

MathsIsFun (http://www.mathsisfun.com)
A rich assortment of materials here for Algebra and Geometry to support learning mathematics.

Maths Online (http://www.univie.ac.at/future.media/moe/galerie.html)
This Austrian website has a variety of materials related to learning algebra. For example: try some examples from *Equations* and from *Functions.*

Skoool (http://lgfl.skoool.co.uk/common.aspx?id=890)
In collaboration with Intel and the Mathematical Association in the UK, Skoool's online *Number Line* is available here; a version can also be downloaded for use offline. The *Mathematical Toolkit* (http://lgfl.skoool.co.uk/keystage4.aspx?id=317# for PC only) can also be downloaded.

TSM-Resources (http://www.tsm-resources.com/mlink.html)
There is an extensive set of links at this UK site, from which many good choices of ICT use can be made, not restricted to Algebra or Geometry.

WaldoMaths (http://www.waldomaths.com)
This UK site contains a large number of Java applets concerned with many aspects of mathematics. For example: Algebra: *Choose some algebra examples from ages 11-16.* Geometry: *Choose some geometry examples from ages 11-16*: *Shape, Space and Measure.*

Geometry Open Book (http://www.mathopenref.com/index.html)
There are many opportunities for pupil to learn by manipulating geometric objects.

Contributing Authors

Kim BESWICK is a Professor in Mathematics Education at the University of Tasmania in Australia and President of the Australian Association of Mathematics Teachers. She taught mathematics and science for 13 years in a number of Tasmanian secondary schools before joining the University of Tasmania in 2000 where she is currently Associate Dean (Research) for the Faculty of Education. She is interested in the beliefs and knowledge that underpin the practice of mathematics teachers and how professional learning can provide a catalyst for change. She supervises research higher degree students in these and other areas of mathematics education. Professor Beswick was co-editor of Australian Primary Mathematics Classroom 2008-2010. She regularly engages in consultancies and research projects involving the design and delivery of education and professional learning for primary and secondary teachers of mathematics across school sectors. She has published more than 80 books, peer reviewed book chapters, journal articles and conference papers.

Margaret BROWN is an Emeritus Professor of Mathematics Education at King's College London. After teaching in primary and secondary schools she joined Chelsea College in London University to train mathematics teachers and undertake research. She has since then directed more than 25 research projects on the learning, teaching and assessment of mathematics in all phases from primary to undergraduate and adult learning. Her particular interest is in the learning and assessment of number concepts. She has been a member of two government-appointed national bodies on the mathematics curriculum, and has received two lifetime awards, two honorary degrees and the Royal Society Kavli medal for her contribution to mathematics education. She is an ex-president of the Mathematical Association and of the British Educational Research Association, an ex-chair of the Joint Mathematical Council of the UK, and ex-Deputy Chair of the Royal Society Education Committee and the Advisory Committee on Mathematics Education.

CHENG Lu Pien is an Assistant Professor in the Mathematics and Mathematics Education Academic Group at the National Institute of Education, Nanyang Technological University, Singapore. She received her PhD in Mathematics Education from the University of Georgia (U.S.) in 2006. She specializes in mathematics education courses for primary school teachers. Her research interests include the professional development of primary school mathematics teachers, tools and processes in mathematics education programs for pre-service teachers. Her research interests also include children's thinking in the mathematics classrooms.

Jeremy HODGEN is a Professor of Mathematics Education at King's College London. He has led many research projects including the ESRC-funded project, Increasing Competence and Confidence in Algebraic and Multiplicative Structure (ICCAMS), which includes a comparison of current Key Stage 3 students' mathematical understandings with those of 30 years ago. He has published widely mathematics education for academic and professional audiences. His research interests include assessment, progression in mathematics, teaching and learning in primary and secondary mathematics, teacher development and international comparisons in mathematics. He is a member of the team coordinating the Targeted Initiative in Science and Mathematics Education in the UK. He is an Editor of the journal, Research in Mathematics Education. Previously, he taught mathematics in both primary and secondary schools.

Berinderjeet KAUR is a Professor of Mathematics Education at the National Institute of Education in Singapore. Her primary research interests are in the area of classroom pedagogy of mathematics teachers and comparative studies in mathematics education. She has been involved in numerous international studies of Mathematics Education and was the Mathematics Consultant to TIMSS 2011. She is also a member of the MEG (Mathematics Expert Group) for PISA 2015. She is the principal investigator (Singapore) of the Learner's Perspective Study (LPS) helmed by Professor David Clarke of the University of Melbourne. As the President of the Association of Mathematics Educators (AME) from 2004-2010, she has also been actively involved in the professional development of mathematics teachers in Singapore and is the founding chairperson of Mathematics Teachers' Conferences that started in 2005. She is also the founding editor of the AME Yearbook series that started in 2009. On Singapore's 41st National Day in 2006, she was awarded the Public Administration Medal by the President of Singapore.

Marian KEMP is now an adjunct Associate Professor in the School of Education at Murdoch University, following on from her appointment in 2011 to the position of Director of Student Life and Learning which she held until the University restructure for 2014. She was a mathematics educator at Murdoch University for about 25 years; firstly in the School of Education and then in the Student Learning Centre where she was the Head of the Centre from 2004 until 2011. Marian provided support for undergraduates in mathematics and statistics across the university. This involved helping students to make appropriate use of scientific and graphics calculators for learning mathematics. She has published in this field with colleagues from Murdoch University and has presented papers and workshops to teachers at conferences nationally and internationally. Over the last 20 years Marian developed programs for improving numeracy across the curriculum. This mainly involved developing critical numeracy tasks, and a series of online numeracy modules. In this area her research has been involved with the development of student strategies for interpreting graphs and tables, including the use of a Five Step Framework. In this field, she has published papers and presented at conferences for both primary and secondary teachers throughout Australia and internationally. In 2007 Marian was awarded a Carrick Institute Award for University Teaching for outstanding contributions to student learning in the development of critical numeracy in tertiary curricula.

Barry KISSANE has been a mathematics educator at Murdoch University in Perth since 1985, except for a period working and studying at the University of Chicago and a recent period as the Dean of the School of Education at Murdoch University. His present teaching responsibilities include mathematics for primary teacher education students and mathematics education for secondary teacher education students. His research interests in mathematics education include numeracy, curriculum development, the use of technology for teaching and learning mathematics and statistics, popular mathematics, teacher education and others. He was written several books and many papers related to the use of graphics calculators in school mathematics, and published papers on other topics, including the use of the Internet and mathematics teacher education. Barry has served as President of the Mathematical Association of Western Australia (MAWA) and as President of the Australian Association of Mathematics Teachers (AAMT). He has been a member of editorial panels of various Australian journals for mathematics teachers for around 30 years, including several years as Editor of The Australian Mathematics Teacher. A regular contributor to conferences for mathematics teachers throughout Australasia, he is an Honorary Life member of both the AAMT and the MAWA.

Dietmar KÜCHEMANN is an affiliate of King's College London. After teaching in several London secondary schools he became a research fellow on the CSMS project at Chelsea College (now part of King's). Part of this work formed the basis of his PhD on children's understanding of generalized arithmetic and also led to a post at the Institute of Education University of London where he taught and sometimes led the PGCE and MA mathematics education courses. He left the IoE to write and publish school mathematics textbooks but returned to work as a research fellow on the Longitudinal Proof Project and the Proof Materials Project. Recently he came back to King's as a research fellow on the ICCAMS project. His main interests are in secondary school students' understanding of algebra and multiplicative reasoning and in developing materials to probe and promote students' and teachers' understanding in these areas and in geometry.

Oh Nam KWON is Professor of Mathematics Education at Seoul National University. She received her PhD in Mathematics from Indiana University in 1992. She received her MA in Mathematics from Seoul National University, her second MA in Education in Mathematics from Indiana University, and her BS from Ewha Womans University. Her earlier professional appointments include Assistant Professor and Associate Professor of Department of Mathematics Education at Ewha Womans University and Visiting Professors of Ohio State University and San Diego State University. She has been involved in more than 30 grants as Principal Investigator and Collaborator. She is Editor of 'The Mathematics Education', 'The SNU Journal of Educational Research', and 'Journal of the Korean School Mathematics Education Society'. She is on Editorial Board for book series "Advances in Mathematics Education" by Springer. She has served as committee member for numerous international (including International Program Committee of ICME-12) and Korean organizations of mathematics education. She is serving as National Committee of Korean Institute of Curriculum and Evaluation. She is a member of OECD/PISA2015 Mathematics Expert Group. She received the Best Teaching Award from Seoul National University in 2009.

Jee Hyun PARK has a PhD from Seoul National University, Korea. Her doctorate was an exploration into how teachers can make computer based assessment for developing students' mathematical process. She has taught middle and high school mathematics for 15 years in Seoul, Korea. She also has been lecturer in a few universities and taught mathematical gifted students. She has been on research teams supported by Korea Foundation for Advance of Science and Creativity. She has been working on programs that develop secondary teachers' professionalism for nearly 10 years. She has been a teaching consultant of school consulting in Seoul Metropolitan of Education for five years. She is recently developing the STEAM (science, technology, engineer, art and mathematics) based curriculum and program for gifted students. Her professional interests are finding ways to increase mathematics literacy for secondary mathematics students and improving teacher preparation program, especially teaching using the technology.

Jung Sook PARK has a Master of Education degree in Mathematics Education and a Doctoral degree in Mathematics Education from Seoul National University. She had taught middle school mathematics for 17 years and has taught high school mathematic for 4 years. Her professional interests are finding ways to increase authentic learning experiences for high school mathematics students and improving teacher preparation programs to include stronger content knowledge for K-12 teachers, especially those that will be teaching middle and high school mathematics.

Yoshinori SHIMIZU, PhD, is a Professor of Mathematics Education at University of Tsukuba, Japan. His primary research interests include international comparative study on mathematics classrooms and assessment of students learning in mathematics. He is the Japanese team leader of the Learner's Perspective Study (LPS), a sixteen countries comparative study on mathematics classrooms. He was a member of Mathematics Expert Group (MEG) for OECD/PISA and has been a member of the Committee for National Assessment of Students Academic Ability in Japan.

TAY Eng Guan is an Associate Professor in the Mathematics and Mathematics Education Academic Group of the National Institute of Education at Nanyang Technological University, Singapore. Dr Tay obtained his PhD in the area of Graph Theory from the National University of Singapore. He has continued his research in Graph Theory and Mathematics Education and has had papers published in international scientific journals in both areas. He is co-author of the books Counting, Introduction to Graph Theory and Making Mathematics

Practical. Dr Tay has taught in Singapore junior colleges and also served a stint in the Ministry of Education.

TOH Pee Choon received his PhD from the National University of Singapore in 2007. He is currently an Assistant Professor at the National Institute of Education, Nanyang Technological University. His main research interest is Number Theory, specifically the theory of partitions, modular forms and elliptic functions. He is also interested in problem solving, the teaching of mathematics at the undergraduate level, as well as the use of technology in teaching.

TOH Tin Lam is an Associate Professor with the Mathematics and Mathematics Education Academic Group, National Institute of Education, Nanyang Technological University, Singapore. He obtained his PhD in Mathematics (Henstock-stochastic integral) from the National University of Singapore. Dr Toh continues to do research in mathematics as well as in mathematics education. He has papers published in international scientific journals in both areas. Dr Toh has taught in junior college in Singapore and was head of the mathematics department at the junior college level before he joined the National Institute of Education.

WONG Khoon Yoong is an Associate Professor in the Mathematics and Mathematics Education Academic Group at the National Institute of Education, Nanyang Technological University, Singapore. He has worked as a mathematics educator in Australia, Brunei Darussalam, Malaysia, and Singapore. He has provided consultancy for education institutes in Chile, Hong Kong, the Philippines, and the United States in mathematics curriculum and teacher education. He has participated in the design and review of the national mathematics curriculum in Malaysia (1970s), Brunei Darussalam (1990s), and Singapore (1980s and 2000s). His research interests cover teacher education, mathematics learning strategies, use of ICT in mathematics instruction, and mathematics problem solving. His recent research projects include Teacher Education and Development Study in Mathematics (TEDS-M), Quality of Teacher Preparation in Secondary Mathematics and Science among APEC

Economies, and Singapore Mathematics Assessment and Pedagogy Project (SMAPP).

Joseph Kai Kow YEO is a Senior Lecturer in the Mathematics and Mathematics Education Academic Group at the National Institute of Education, Nanyang Technological University, Singapore.   As a mathematics educator, he is involved in training pre-service and in-service mathematics teachers at primary and secondary levels and has also conducted numerous professional development courses for teachers in Singapore and overseas.   Before joining the National Institute of Education in 2000, he held the post of Vice Principal and Head of Mathematics Department in secondary schools.   His research interests include mathematical problem solving in the primary and secondary levels, mathematics pedagogical content knowledge of teachers, mathematics teaching in primary schools and mathematics anxiety.

ZHAO Dongsheng is an Associate Professor in the Mathematics and Mathematics Education Academic Group at the National Institute of Education, Nanyang Technological University, Singapore. He is a mathematician with research interests in general topology, order structures (posets, domain theory, special types of lattices), and real analysis (generalized Riemann Integral, Baire class one functions). He was also a member and Co-Principal Investigator of the mathematics education project SMAPP (Singapore Mathematics Assessment and Pedagogy Project) between 2008 and 2012. He received his BSc degree in Shaanxi Normal University and PhD from the University of Cambridge (UK).

Printed in the United States
By Bookmasters